Taliby Dos CAMARA

VIROLOGY COURSE MANUAL

Taliby Dos CAMARA

VIROLOGY COURSE MANUAL

First edition

ScienciaScripts

Imprint

Any brand names and product names mentioned in this book are subject to trademark, brand or patent protection and are trademarks or registered trademarks of their respective holders. The use of brand names, product names, common names, trade names, product descriptions etc. even without a particular marking in this work is in no way to be construed to mean that such names may be regarded as unrestricted in respect of trademark and brand protection legislation and could thus be used by anyone.

Cover image: www.ingimage.com

This book is a translation from the original published under ISBN 978-620-6-70097-5.

Publisher:
Sciencia Scripts
is a trademark of
Dodo Books Indian Ocean Ltd. and OmniScriptum S.R.L publishing group

120 High Road, East Finchley, London, N2 9ED, United Kingdom
Str. Armeneasca 28/1, office 1, Chisinau MD-2012, Republic of Moldova, Europe
Printed at: see last page
ISBN: 978-620-7-63033-2

Copyright © Taliby Dos CAMARA
Copyright © 2024 Dodo Books Indian Ocean Ltd. and OmniScriptum S.R.L publishing group

Contents

Dedication

I dedicate this book to my dear wife and children:
- *Djenabou CAMARA*
- *Fatoumata Taliby*
- *Mariam Taliby*
- *Mohamed Taliby*
- *Bintou Taliby*
- *My nephew Soriba CAMARA   for their   unconditional love on a daily basis*

ACKNOWLEDGEMENTS

To God Almighty, Most Merciful and Most Merciful, for giving me life, health, courage, energy and so many other good things that have enabled me to reach this important stage in my destiny. May He bless my efforts and make my faith grow now and forever. Amina !

To all my fellow Teachers and Researchers in Guinea and elsewhere (SOACHlM and CAMES).

To all my pupils at College 1 Donka in 1998, Lycee de Kipe, Groupe Scolaire Koumba Diawara and students at the Gamal Abdel Nasser University of Conakry (UGANC), the Mahatma Gandhi University (UMG) and the Institut de Recherche en Biologie Appliquée de Guinee (IRBAG) for all the time spent together in the precious field of Biological Sciences.

To the Dean of the Faculty of Sciences of the UGANC, Pr Ousmane BARRY, to the Head of the Department of Biology, Pr Demba MAGASSOUBA, to the former Vice-Dean in charge of Research, Pr Lipoli-pe KOLIE, to the Vice-Deans, Dr Ce CAMARA and Dr Ibrahima Sory Leon CAMARA, to the Directors of the Programmes of Biology, M. Lansana BANGOURA, of Biochemistry, M. Cheick Ahmed Tidiane CAMARA, of Medical Biology, Dr Ahmadou Sadio DIALLO for their support as well as to all the others. Lansana BANGOURA, of Biochemistry, Mr Cheick Ahmed Tidiane CAMARA, of Medical Biology, Dr Ahmadou Sadio DIALLO for their support, as well as all the others not mentioned. May they all find here the expression of my highest consideration.

To the Director of Advanced Studies and former Minister of Higher Education and Scientific Research, Pr Aboubacar Oumar BANGOURA, for the quality reforms carried out in favour of the ASP.

To the former and current managers of lRBAG, Pr Mamadou Yero BOlRO, Pr Mamadou Cellou BALDE, Pr Siba KALIVOGUI, Pr Sanaba BOUMBALY, Pr Mohamed Sahar TRAORE, Dr Aissatou BOIRO for the good collaboration.

To the Founder of Mahatma Gandhi University, Mr Mamadou Mouctar SAVANE, to the Rector, Dr Lansana CAMARA, to the Vice-Rector, Dr Yao AGBENO and to the CFO, Mr CISSE, for their good and frank collaboration.

To my dear Masters and Mentors at CAMES, Pr Mohamed CISSE, Dean of the Faculty of Health Sciences and Techniques and Director General of the CHU Donka National Hospital and Pr Telly SY, Head of the Chair of Gynecology and CAMES Focal Point at the Gamal Abdel Nasser University in Conakry.

To my Dear Masters, Dr Thierno Ibrahima DIALLO and Dr Mamadou SANGARE for their attention and support in all my projects.

A special mention to my great sister Mrs TOURE Khalifatou, her children and her late husband Fode Lamine TOURE. The latter had been a former school inspector, former Governor of the Nzerekore region, former Administrator of the University of Conakry, former Chief of Staff of the Ministry of National Education, writer, poet essayist, for having been my mentor and for having inspired me. May Allah's eternal Paradise be his home. Bring it on!

A special hello to El-hadj Oumar BAH, Pr Mariam BEAVOGUI, M. Mohamed MAGASSOUBA, Dr Mohamed Ansoumane CAMARA, Dr. Issa Samaya SYLLA, Dr Mohamed Kerfalla CAMARA, Dr Mamady Balla Junior CAMARA, Mme Sira SYLLA, Dr Sekou DIAKITE, Dr Abdoulaye KABA, Pr Alia DIABY, M. Sekou MONDEKENO, M. Bangaly SYLLA, etc. for their sincere friendship!

Mr Souaibou SOW, Head of the Medical Biology Laboratory at Mahatma Gandhi University, for his contribution to this work!

Finally, I would like to thank all those who have contributed, in one way or another, through their advice and encouragement. I would like to express my deep gratitude to them all.

PREFACE

Virology is the study of viruses and associated infectious agents. This book of General and Special Virology is intended for students of biology, medicine, veterinary medicine and laboratory technology. It is a rich document in terms of scientific information, as it helps to understand viruses in relation to infected organisms. In writing this Virology manual, the author, Dr Taliby Dos Camara, as a microbiologist, wishes to make his contribution to the knowledge and control of viruses of medical interest at a time when the world is experiencing daily viral epidemics such as avian influenza, Zika, Ebola, Lassa, Covid-19, foot-and-mouth disease, and so on.

This document is divided into three main parts. The first part deals with General Virology, which studies all the properties of viruses. The second part is devoted to Special Virology, which deals with nine viruses and the nine diseases they cause. The third is devoted to methods of diagnosing viral diseases. Consolidation questions to enable learners to self-assess their level of learning from the course.

In this manual, the author aims to describe the structure of viruses, the mechanisms that enable them to infect cells and to take advantage of cellular mechanisms to reproduce. This study also covers the diseases caused by viruses, the techniques for isolating and cultivating them, and their use in research and therapy. Virology is generally considered to be a branch of microbiology or pathology.

In this book, the Author has developed the properties of viruses, namely the structure of the viral particle, virus replication, their pathogenies, methods of diagnosing viral diseases, the ways in which naked viruses penetrate host cells, modes of penetration of enveloped viruses, gene expression according to the nature and polarity of the nucleic acid, compartmentalisation of eucaryotic cells and virus internalisation, the replication cycle, plant viruses and gene expression in phytoviruses.

This book on virology is a veritable goldmine, and a must-have!

Pr AbdoulayeMAKANERA, Lecturer-Researcher, Bacteriology-Virology, at the Gamal Abdel Nasser University in Conakry and Head of the Biomedical Analysis Laboratory at HASGUI.

CHAPTER I: GENERAL VIROLOGY

1. INTRODUCTION

The ancients were familiar with diseases such as rabies, measles and smallpox, but they did not understand their nature. The smallpox epidemics of 165 to 180 and 251 to 266 AD are probably partly responsible for the weakening of the Roman Empire.

The word **virus is** often confusing because it can refer to several entities or concepts. Historically, it was used to refer to infectious agents that could be transmitted regardless of their nature, since this was unknown. Nowadays, in popular language, it refers more specifically to certain infectious agents, such as the influenza virus, without referring to a specific biological structure.

In its current, broadest scientific acceptance, a virus is an intracellular parasitic biological entity that is required for replication and propagation. In this acceptation of the term, "virus" means the cellular machinery set up by the viral programme in one or more structures of the cell, necessary for its replication. The viral particle, whose role is to carry the viral genome from one cell to another and from one host to another, is just one of the stages in the replication cycle of the "virus".

However, the term virus is often still used to refer to the extra-cellular viral particle. The term **"virion"** removes this ambiguity by designating only the viral particle, the vehicle of the viral genome. In the context of virus structure, we are indeed interested in the structure of the virion, which is not, however, an inert structure. On the contrary, this structure has the function of ensuring the first stages of the viral replication cycle, i.e. recognition of the target cells and penetration of the viral genetic material into them.

Viruses are generally much smaller than bacteria, 1 micron (10^6 m) for Staphylococcus. Their size varies from around ten nanometres for the smallest to several hundred nanometres (10^{-9} m) for the largest. Recently, giant viruses have been discovered, such as Mimivirus, an amoeba virus, which is the same size as mycoplasma. In general, however, viruses cannot be observed using light microscopy, and electron microscopy is required.

Recent technical developments, such as cryotomo-electron microscopy with 3D structure reconstruction using the computing power of computers, combined with protein crystallography data, have made it possible to resolve certain viral structures to within a few Angstrom. The images obtained for three viruses belonging to different families show that the viruses have more or less spherical globular structures, composed of several layers, with the viral genome located in the centre of the virus.

More or less complete, the structure of viruses can comprise a protein shell, the capsid, which contains the viral genome, and an envelope made of a bi-layer of phospholipids into which glycoproteins, known as envelope proteins, are inserted.

The structure of a virion can be reduced to its simplest expression: a capsid containing the viral genome, which may be DNA or RNA. This is the case, for example, with adenoviruses or Rotaviruses. These are known as naked viruses. Alternatively, a virion can be enveloped if the capsid is surrounded by a phospholipid bilayer with its proteins. In general, naked viruses are much more resistant in the external environment than enveloped viruses, whose envelope is more easily damaged by solvents, detergents and the physical conditions of the environment, humidity, drought, etc.

Many viral capsids are icosahedrons, the geometric characteristics of which are shown in these diagrams with faces, vertices and axes of symmetry. A football is a frequently encountered icosahedral with alternating pentons (five sides) that make up the vertices, symmetry axis 5, and hexons (six sides) that make up the faces. It is this alternation that curves the surface to create a sphere. The basic elements of these structures are capsomeres, made up of one or more proteins.

Several capsomers are needed to make a face. Their number can vary as shown here, but always with the same rules of triangulation. Papillomaviruses are an example of a simple naked virus made up of 72 capsomeres. You can look for the Pentons and Hexons on the reconstructed image and spot the triangulation.

For adenoviruses, Pentons and Hexons are different. Pentons also have an extension, the fibres.

Rotavirus capsid proteins are more diverse, producing more varied patterns, but the same geometry is always found.

It should be borne in mind that capsids have two roles to play. Firstly, they protect the viral genome which is enclosed in their cavity and, secondly, they must also bind to receptors on the surface of the target cells. This recognition is achieved by protein motifs exposed directly on the capsids or which demask after the action of proteases, digestive proteases for example. It is easy to imagine that antibodies specifically recognising these viral protein motifs could block the virus from attaching to the target cell. These are known as **neutralising antibodies: Helicoidal capsids.**

Another common form of capsid is the helical capsid, which consists of a stack of capsomeres along the viral genome in a helical arrangement. These capsids can be naked or enveloped, as in the examples shown here.

Finally, capsids can have much more complex structures. This is the case, for example, with Poxviruses

In the case of enveloped viruses, the capsid is surrounded by an envelope. This is made up of a phospholipid bilayer like those of plasma cell membranes or those of the endoplasmic reticulum. The envelope is more or less loose around the capsid, as in the photo on the right. Viral proteins are always associated with the envelope via transmembrane domains. The space between the capsid and the envelope is occupied by matrix proteins.

Envelope glycoproteins are complex structures which play major roles in the attachment of the virus to target cells and its internalisation into them. They are classified into three different types. Examples of glycoprotein types are shown here. These proteins are homo- or hetero-trimeric plugs perpendicular to the viral envelope.

This is the case with the H protein of the influenza virus, which assembles in the form of a homo trimer (left). The conformation of this trimer varies according to pH. At neutral pH, its structure enables it to recognise its receptor on the surface of cells in the upper airways. After binding to its receptor, the virus is internalized by endocytosis. Acidification in the endosome induces a change in the conformation of the trimer, which unmasks hydrophobic domains that bind to the endosome membrane, resulting in fusion of the viral envelope with the endosome membrane. In this book, we return to influenza.

Another example of a type 1 envelope protein is that of the human immunodeficiency virus, HIV. This virus has a structure consisting of a capsid containing the two copies of the viral genome, a matrix protein lining the inside of the envelope, and an envelope with heterotrimeres of two viral glycoproteins, gp120 and gp41.

It is easy to imagine that these proteins, whose conformation changes as a function of interactions with receptors, are difficult to recognise in their native state by antibodies, making it easier for viruses to escape the immune response.

Viruses are grouped together on the basis of a set of characteristics that they share to a greater or lesser extent. These characteristics include, of course, the characteristics of the structure of the virion and the nature of their genome, but also, more importantly, the biological characteristics of the "virus" in the broadest sense of the term. In other words, the cellular and host tropism of the virus, as well as the characteristics of the viral replication cycle within the cell.

2. Definition of viruses

Viruses are particulate, infectious, subcellular biological objects, endowed with genetic continuity (replication of genetic material) and great evolutionary capacity, consisting of at least one nucleic acid (DNA or RNA) and proteins; they depend on living cells to be replicated and to do this, they are capable of profoundly and/or durably disrupting the genetic information of the cells they infect (Chastel, 1992).

In other words, viruses are cell-free organisms, absolute intracellular parasites, made up of a single nucleic acid (DNA or RNA) and proteins, with cubic or helical symmetry, which can be naked or enveloped, without metabolism or autonomous multiplication.

3. History of virology

Viral diseases such as smallpox and rabies have been known since ancient times. At the end of the 18th century, ***Edward Jenner*** developed the inoculation of "cowpox" or bovine smallpox, which provided good protection against smallpox. The name vaccination derives from the word "vacca", cow, and is applied by

Louis Pasteur in the following century to the rabies vaccine, obtained by attenuating infectious material passed on to animals. The first experiment indicating the involvement of an ultrafiltrable agent, smaller than bacteria, in certain infectious diseases, was the transmission of tobacco mosaic by ***Dmitrii Ivanovski*** from plant filtrates in 1892. But it was only 6 years later that ***Martinus Beijerinck*** understood the consequences of this observation by repeating it. He spoke of "*contagium vivum fluidum*". In the late nineteenth and early twentieth centuries, many viruses were rapidly discovered in animals and humans. It was also discovered that certain viruses could infect bacteria. These were called "bacteriophages" by ***Felix d'Herelle***. Viruses were visualised for the first time using an electron microscope in 1939, and after 1948 cell culture techniques enabled new viruses to be isolated and characterised. In 1979, the *World Health Organisation certified the global eradication of smallpox*. It was the first great triumph of medicine, and of vaccination in particular. When **AIDS** was described in 1980, it would take just three years to discover its causal virus, HIV. Techniques continued to evolve, the most recent advance being the development of the polymerase chain reaction (PCR) by ***Kary Mullis*** in 1985, and new viruses continued to be discovered every year.

4. General characteristics of viruses

Viruses are replicating elements that are much smaller than bacteria, and the larger ones are barely visible under an optical microscope. Their genome can be composed of either RNA or DNA. Viruses are highly dependent on cellular metabolism. In the cell they infect, they replicate their genome and protein components separately; these are then assembled, giving thousands of particles in one generation. Viruses will specifically recognise one or a few cell types and are therefore quite host organism specific.

4.1 Viruses are very small

The main characteristic of viruses, to which we owe their discovery, is their ability to pass through filters that are impermeable to bacteria. While the largest viruses infecting humans, the Poxviridae, are between 250 and 300 nm in size, the smallest, the Parvoviridae, are just 20 nm. Size is not an absolute criterion, however, and the Mimiviruses described in 2003 in the amoebae Acantamoeba polyphaga are the size of small bacteria such as rickettsiae (+/- 1gm).

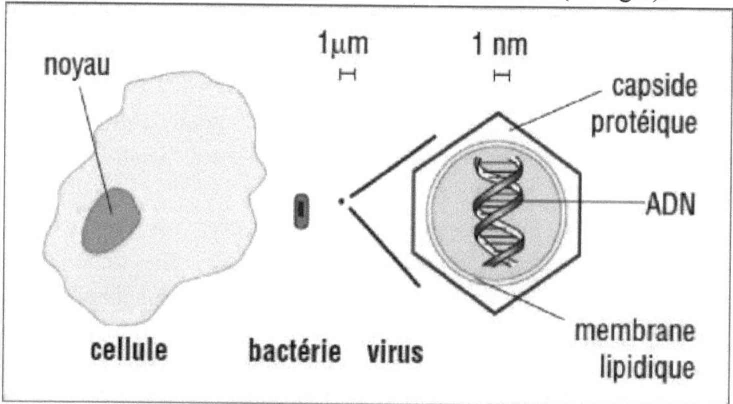

Figure 1: Animal cells, bacteria and viruses

4.2 Viruses replicate themselves

Another basic characteristic of viruses is that they replicate. For example, infection by tobacco mosaic virus can be propagated indefinitely from plant to plant, even if the inoculum is diluted on each pass. This distinguishes viruses from toxins, which lose their toxicity through dilution.

4.1.1 Each viral particle contains only one type of nucleic acid

Viruses are essentially made up of a molecule carrying genetic information. This can be present in the form of either DNA or RNA in the viral particles. Viruses are therefore separated according to their DNA or RNA composition. Some viruses have an intermediary of their genome in a different form during replication: retroviruses, which are RNA viruses, will be retrotranscribed into DNA in the host cell, and it is from this "proviral" DNA that the new RNA genomic strands will be formed. In a similar way, Hepadnaviridae, such as the human hepatitis B virus or certain plant viruses, are DNA viruses which pass through an RNA intermediary to form the new DNA strands.

4.1.2 Viruses are replicas assembled from their components

This notion is well illustrated by *E. Ellis and M. Delbruck*'s experiment with bacteriophage X. (See also the section on the viral cycle) A suspension of bacteriophage X is mixed with bacteria in a ratio of 10:1. During infection, there is first of all a phase in which no infectious virus can be recovered from the infected cell: this is because, during infection, the genome has been released from the capsid (decapsidation). There is therefore no longer a complete, infectious virion. This phase is known as the "eclipse phase". Next, the viral genome is transcribed and provides the proteins encoded by the virus. The genome is also replicated to give rise to new copies of the viral genome. These replicated genomes combine with the structural proteins of the virus (assembly) to form new infectious virions.

During this phase, known as the maturation phase, new infectious viral particles are assembled in the cell from their components. This contrasts with the replication cycle of a cell or bacterium, during which the daughter cell is not formed de novo by a process of assembling components of the parent cell, but is formed by splitting the parent cell.

4.1.3 Viruses are strictly dependent on the metabolism of a cell

The fact that viruses have no ribosomes and have to be assembled from scattered elements makes them dependent on a favourable environment, which is that of a cell. These cells may be of different types - bacteria, algae, plants or animals - and will in some way be at the service of the virus. It is therefore obvious that viruses cannot replicate in an amorphous medium, such as a bacteriological culture broth. Some bacteria are also intracellular, but unlike viruses they have most of the elements needed for their metabolism and replication, particularly ribosomes.

4.1.4 Viruses are specific to cells and organisms

All living organisms are susceptible to infection by viruses, but it is not the same viruses that infect different organisms. So plant viruses will not generally infect animals, and viruses from one species of plant will not necessarily infect other species of plant.

This species barrier is not absolute and we can see, for example, that certain animals can share viruses with humans and with each other. Within an organism, viruses will be selective for certain types of cell. This specificity is largely due to the specific receptors on the cell surface that allow the viruses to bind and enter. For example, AIDS viruses recognise certain immune system cells carrying the CD4 molecule on their surface.

5. Structure of the viral particle

The study of viral structure has led to a better understanding of viruses and how they function. By understanding how the virion is constructed, we can gain a better understanding of several essential stages in the viral cycle, such as attachment, penetration, decapsidation, virus assembly and exit. In addition to the functions associated with virus attachment, penetration and exit, the viral capsid undoubtedly has a virus protection function, particularly in the case of viruses transmitted in aerosol form (influenza virus) or mechanically to plants (tobacco mosaic virus). In recent years, we have also realised that the capsid can be a dynamic structure.

Precise knowledge of viral structure is of major interest in vaccine research and nanotechnology: what could be more fascinating than the ability of a virus to encapsulate a nucleic acid molecule in a specific way, in the complex environment of a cell!

Researchers such as Crick and Watson and Klug have been fascinated by these issues, using techniques such as X-ray diffraction and electron microscopy to decipher the architecture of many known viruses and the way they are assembled.

Systematically, the virus is composed of a genome and a capsid, a shell that surrounds the viral nucleic acid. This capsid is formed by the assembly of repetitive protein subunits, sometimes called capsomeres. The assembly formed by the capsid and the viral nucleic acid is called the nucleocapsid.

Electron microscopy has revealed two main types of capsidal structure: elongated particles and spherical particles.

In addition to the capsid and viral nucleic acid, some viruses are surrounded by a lipid envelope, sometimes called the peplos (coat): these are known as "enveloped" viruses. In the absence of an envelope, however, they are referred to as "naked" viruses.

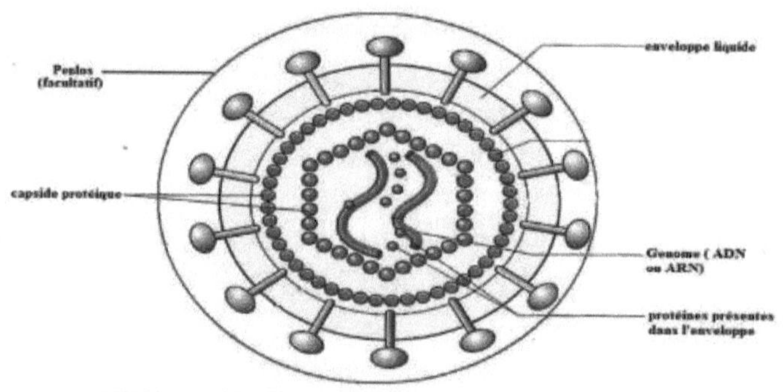

Figure 2: Structure of an envelope virus

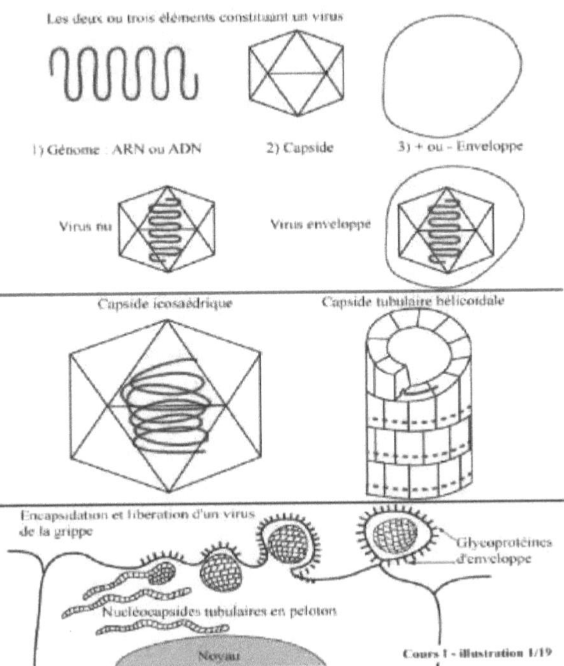

Figure 3: Components of viral particles

5.1 Virion components

5.1.1. Viral genome

A virus usually consists of a genome made up of one or more strands of deoxyribonucleic or ribonucleic acid, in linear or circular form. A distinction is made between single-stranded and double-stranded RNA and DNA, RNA of positive or negative polarity or ambisens.

Example of a typical capsid protein structure It is made up of 150 to 200 amino acids arranged in eight antiparallel beta sheets to form a "trapezoidal" or barrel structure. Viral RNAs can be capped, covalently associated with a protective protein, end with a polyadenylated sequence or with a pseudotRNA end containing a pseudo-node.

5.1.2. Capsid proteins

Capsid proteins are remarkable proteins. They are capable of polymerising by self-assembly to form the complex structures that are viral capsids. In some cases, they can also interact specifically with viral nucleic acids. Some viral capsid proteins have been studied in detail, such as the capsid protein of VMT-TMV. Typical icosahedral viral proteins have a characteristic structure, consisting of 150 to 200 amino acids arranged in eight antiparallel beta-sheets to form a 'trapezoidal' or barrel structure.

5.1.3. Matrix protein

Some viruses, such as Retroviruses, have matrix proteins that enable the nucleocapsid to bind to the envelope via a transmembrane anchoring domain. These proteins are generally not glycosylated. However, they often make a significant contribution to the mass of the viral particle.

In Herpesviridae, the proteins located between the membrane and the capsid are called "tegument".

5.1.4. Viral envelopes

Most plant viruses are naked viruses, i.e. non-enveloped, with the exception of Rhabdoviruses and Tospoviruses. This can no doubt be explained by the marked difference between the walls of plant cells and those of animal cells. On the other hand, many animal and insect viruses have an enveloped capsid structure. Bacteriophages, on the other hand, can be naked, enveloped or have a membrane inside the capsid, enveloping the genome (as in the case of the Tectiviridae).

The envelope plays a key role in the attachment of the virus to the target cell, via membrane glycoproteins specific to cellular receptors. A typical example of a membrane glycoprotein is the hemagglutinin of the Influenza virus.

The viral envelope is made up of glycoproteins of viral origin, sometimes called spicules. Some of these have a transmembrane anchoring domain, and are often highly glycosylated at their extra-viral end. In some cases, the glycoprotein is more than 75% carbohydrate by weight. These proteins are generally remarkable antigens, and have several functions: for example, hemagglutinin acts as an elicitor (binding to a cell receptor) and enables membrane fusion. Its carbohydrate-binding properties are exploited in the hemagglutination and hemagglutination inhibition assays.

The viral envelope often also enables the infection to be initiated, by allowing the nucleocapsid to be released into the cell cytoplasm. In the other direction, budding allows the virus to leave the cell without causing complete cell lysis, thus avoiding subjecting the host to too much pressure.

Viral envelopes also contain membrane transport proteins, comprising several hydrophobic transmembrane domains. These proteins ensure exchanges between the virion and the outside world and play an essential role in the biochemical maturation of viral particles. The M2 protein of the Influenza virus is an example of this type of protein (see diagram of an influenza virus with glycoproteins forming spicules and M2 proteins forming transmembrane channels).

5.1.5. Anti-receptors

The envelope is the support for the determinants of virus-host cell recognition in enveloped viruses. These glycoproteins (spicules) enable the virus to recognise the target cell via a cellular receptor, and are therefore sometimes referred to as anti-receptors. We now know more and more about cellular receptors and their viral anti-receptors. We have thus been able to describe super families or groups of characteristic receptors.

6. Viruses with helical symmetry

Two main types of viral structure have been identified: elongated viruses with a helical structure, either naked or enveloped, and quasi-spherical viruses with an icosahedral structure.

6.1 Naked viruses with helical symmetry

Elongated viruses therefore have helicoidally symmetrical particles. When the capsid of this type of virus is not enveloped, it is referred to as a "naked" virus. These are essentially plant viruses and some bacteriophages. These viruses can encapsulate a nucleic acid whose size is not limited a priori. The virus with the best known helical symmetry is the tobacco mosaic virus (VMT- TMV).

6.2 Enveloped viruses with helical symmetry

Other viruses with helical symmetry form elongated but 'flexuous' particles. In this case, protein-protein interactions are weaker than in the case of rigid viruses. Potato viruses X and Y (Potato virus X - PVX, Potato virus Y - PVY) are examples of this type of virus.

Several viruses have helical symmetry while being enveloped, including all animal and human viruses with helical symmetry. Myxoviruses (Orthomyxoviruses and Paramyxoviruses) and Rhabdoviruses are the main viruses with this particular structural form. The viral nucleic acid is surrounded by a capsid to form a flexible nucleocapsid, coiled more or less regularly in the virion, whose envelope is made up of glycosylated proteins and lipids.

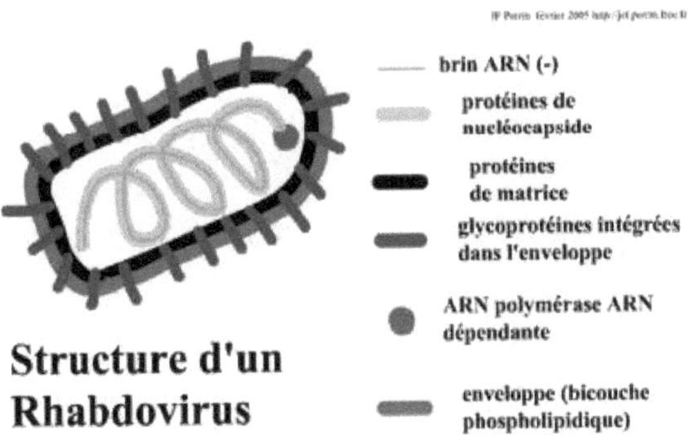

Figure 4: structure of the rabies virus

Vesicular stomatitis Indiana virus (VSIV), rabies virus (RABV) and alfalfa mosaic virus (VMl-AMV) (Figure) have a characteristic bullet shape. A major protein, the n protein, surrounds the viral ribonucleic acid. A matrix protein provides the link between this nucleocapsid and the envelope enclosing the glycoproteic spicules.

The influenza virus has a complex architecture, comprising up to eight distinct nucleocapsids within a complex lipoproteic envelope, covered with spicules made up of two glycoproteins of viral origin, hemagglutinin and neuraminidase, which play an important role as antigenic determinants (see opposite the diagram of an influenza virus nucleocapsid with a simplified representation of neuraminidase, a simplified representation of hemagglutinin and a simplified representation of the M2 protein).

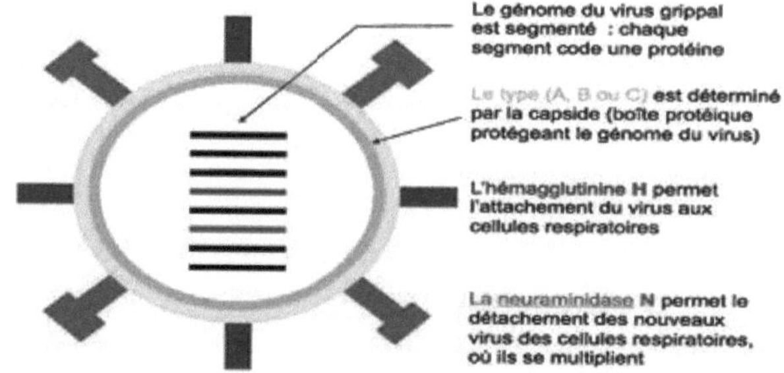

Figure 5: Structure of the influenza virus

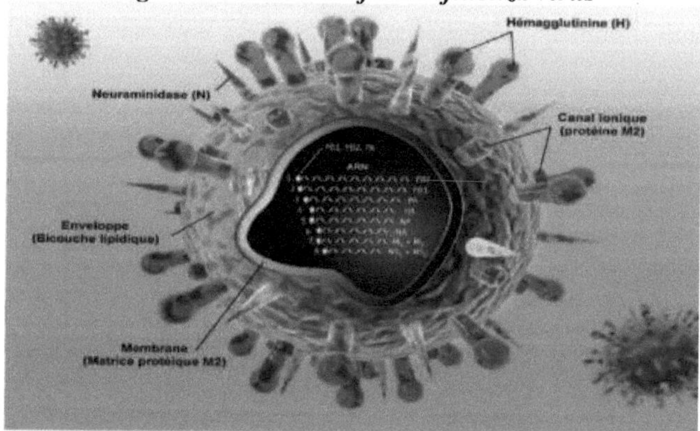

Figure 6: Structure of the influenza virus

7. Viruses with icosahedral symmetry

The architecture of small, spherical viruses has long intrigued scientists, as it raises a series of interesting questions: how is a virus whose genome is sometimes limited to a few thousand nucleotide bases capable of producing a complex capsid composed of several hundred proteins? How are the protein subunits able to interact with each other? What is the size of the nucleic acid that can be encapsidated in this type of structure?

It is possible to arrange symmetrically identical protein subunits to create a quasi-spherical structure. In theory, it is possible to construct a tetrahedral (four triangular faces), a cube (six square faces), an octahedral (eight triangular faces), a dodecahedral (12 pentagonal faces) and an icosahedral, a quasi-spherical shape with 20 triangular faces (Figure).

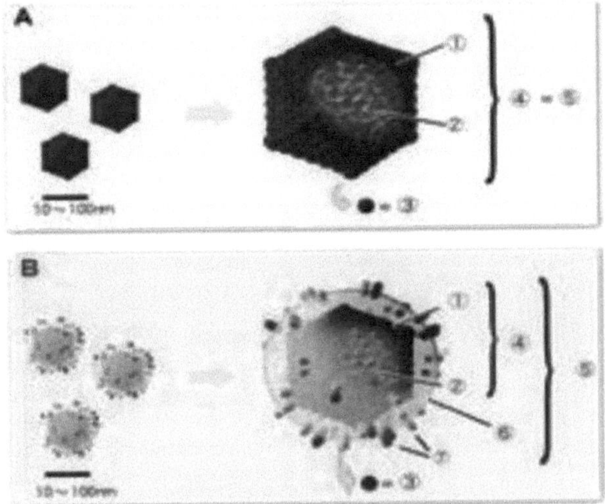

Figure 7: Icosahedral structure

This structure corresponds to data obtained in the early 1960s for a series of small, spherical-looking viruses. It is more economical for the virus to encapsulate its genome in a capsid made up of several identical repeated subunits, than by using fewer but larger subunits. Moreover, it is unlikely that a tetrahedron could contain the genome of an entire virus, and if even a virus managed such a feat, it is likely that the capsid thus created would not fulfil its primary role: protecting the viral genome! It is also important to stress that the size of the genome that can be encapsidated in an icosahedral-type virus is limited compared with helicoidal capsid viruses, which are in principle unlimited in length...

The icosahedron is a regular polyhedron with three axes of symmetry, 12 vertices, 20 faces that are equilateral triangles and 30 edges.

For a given virus, the number of proteins needed to assemble an icosahedral capsid is indicated by the triangulation number T : t X 60 proteins are needed to build the capsid.

In the case of the smallest known viruses, such as phage 0174 (Microviridae), the number is equal to 1. Close examination of electron micrographs shows that the number and quantity of structures apparent on the surface of virions often does not correspond to a multiple of 60. This shows that the proteins on the surface of the capsid are not necessarily grouped together in the equilateral triangles that form the pseudo-icosahedral, but may be distributed in a different way. These groups of proteins are known as capsomeres.

8. Viruses with complex architecture

A number of viruses construct their capsids in a way that does not correspond to helical or icosahedral standards. For example, phages in the t series have a binary structure, involving both helical and icosahedral elements (Figure 8: Bacteriophage).

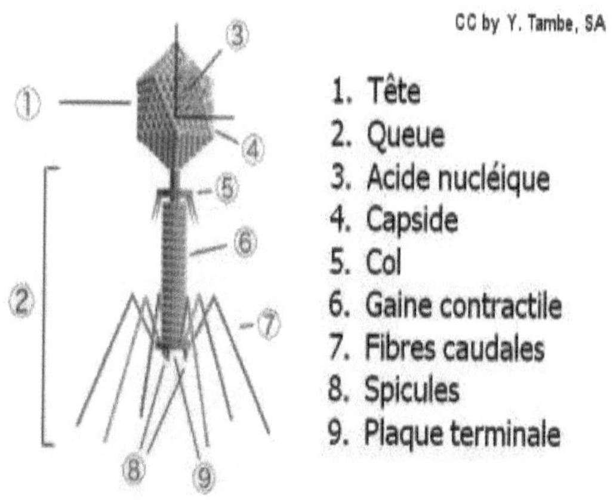

Figure 8: Structure of a Bacteria virus

Criteria for distinguishing viruses

Essentially based on the type of viral genome, the replicative strategy or the structure of the particle, the taxonomy of viruses occupies a special place. The concept of a viral species, in the absence of sexual reproduction, is quite specific. This section also provides access to resources for finding the name of a given virus, such as the database of the International Committee for Taxonomy of Viruses (ICTV).

Systematics is the study of the types and diversity of organisms and the relationships between them. This fundamental discipline combines the classification of organisms or taxonomy, the nomenclature used to name them and the analyses linked to the study of the revolution of organisms and their phylogeny. However, it is widely accepted that the origin of viruses is probably multiple, and that the multiple recombinations and reassortments between viral genomes imply the existence of chemical organisms and polyphyletic genomes...

9. Taxonomic notion of viruses

From a scientific point of view, it is vital to have a clear understanding of the names of viruses. Nevertheless, the classification of viruses is original compared with the Latin binomial nomenclature usually used for living organisms.

In practical terms, the need to divide the world of viruses into easily identifiable and universally accepted entities is the justification for developing a classification system. Initiated in 1966 at a congress in Moscow, a committee was set up within the virology division of the International Union of Microbiological Societies (ICTV, International Committee for taxonomy of viruses), with the task of proposing a taxonomic scheme for all viruses, whether animal (vertebrates, invertebrates, protozoa), plant, algal, fungal or bacterial...

9.1 Virus classification criteria

The first criteria used to distinguish viruses were clinical or pathological (e.g. tobacco mosaic virus named according to the disease produced), as well as elements of an ecological nature or linked to the mode of transmission and the vector. These elements are, of course, still used, but these functional groupings, which can be used for practical reasons, are not taken into account in the actual taxonomy of viruses (e.g. hepatitis, caused by viruses that are very different from one another).

With the development of electron microscopy, the morphology of viral particles became the primary criterion used.

Although most viruses could be observed using microscopy, other factors were soon taken into account, such as stability (at different pH levels, detergents, temperatures, etc.) and antigenicity.

But today, the essential criteria used in taxonomy are :
- the type of viral genome and its organisation
- viral replication strategy (see table below for diagrams of the different forms of RNA and DNA viruses)
- the structure of the viral particle.

The most practical classification is probably that based on the type of nucleic acid (DNA or RNA) and its mode of expression.

9.1.1 Classification of the LHT system

It is based on four criteria, namely the type of viral genome (DNA or RNA), the type of symmetry, the presence or absence of an envelope and the number of capsomeres. This classification does not include hepatitis viruses (which are not classified).

R=RNA nucleic acid D=DNA	Summary H- HelicoPidale C=cubic	Bare (N) or Envelope (E)	Groups	Propeller diameter (nm)	Number of capsomeres	Types
		N	Virus Vegetables	17		V.tobacco mosaic (TMV)

R	H	E	Orthomyxo-viridae Paramyxo-viridae Rhabdo- viridae	9 18 18		M.influenzae A.B Paramyxovirus1,2,3,4 Mumps V V of measles Respiratory syncytial virus V rabies
		N	Picorna viridae Reoviridae		32	Enterovirus : Poliovirus 1,2,3 ECHO virus 1,2..31 V. Coxsackie A, B
					92	Reovirus 1,2,3
D	C	E	Togaviridae			V of yellow fever Encephalitis Measles
		N	Papova-Viridae Adeno viridae		72 252	Polyoma wart, Butterfly SV 40 Adenovirus 12 35
		E	Herpes Viridae		162	Herpes Shingles, chickenpox Inclusion Cytomegalovirus Epstein-Barr virus
	H	E	Poxviridae	910		Smallpox, vaccine

NB : Limitations : Hepadnaviridae have not been classified.

9.1.2 The Baltimore classification

This classification, used today as a basis by the ICTV, was initially proposed by David Baltimore, winner of the Nobel Prize for Medicine in 1975.

Classifying viruses according to their genome means that within a given category they all behave in the same way, which gives some indication of whether further research is needed. The classification is :

DNA viruses: Groups I and II :

- Group I: Double-stranded DNA viruses (Adenovirus, Herpesvirus, Poxvirus) - Group II: Single-stranded DNA viruses: DNA with (+) polarity (Parvovirus)

RNA viruses: Groups III, IV and V :

• Group III: Double-stranded RNA viruses (Reovirus)

• Group IV: Polar-positive single-stranded RNA viruses: polar-positive (+) RNA (Picornavirus, Togavirus, Coronavirus)

• Group V: Negative-polarity single-stranded RNA viruses: negative-polarity (-) RNA (Orthomyxovirus, Rhabdovirus)

Reverse transcription viruses: **Groups VI and VII :**

• Group VI: Retroviruses with single-stranded RNA: RNA with positive polarity (+) with intermediate DNA in the life cycle (Retroviruses)

• Group VII: Double-stranded DNA pararetroviruses (Hepadnavirus)

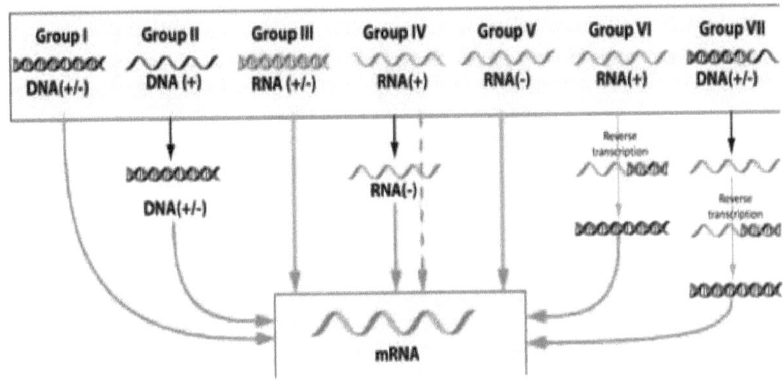

Figure 9: Classification of viruses according to David Baltimore

DNA viruses

Group I: Double-stranded DNA viruses

This type of virus usually has to enter the host cell nucleus before it can replicate. In addition, these viruses need the host cell's DNA polymerase to replicate the viral genome and are therefore highly dependent on the cell cycle. Propagation of the infection and production of new viruses requires the cell to be in a replication phase, as this is when the cell's polymerases are active. The virus can induce forced cell division, and when this occurs chronically, it can lead to cell transformation and, ultimately, cancer. Known examples include the Herpesviridae, Adenoviridae and Papovaviridae.

There is only one well-studied example of a group I virus not replicating in the nucleus, and that is the poxvirus family, a group of highly pathogenic viruses that infect vertebrates, one of whose representatives is the smallpox virus.

Group II: Single-stranded DNA viruses

The viruses that fall into this category include certain infectious agents that have not been as well studied, but are still closely linked to vertebrates. Two examples are the Circoviridae and the Parvoviridae. They replicate in the nucleus and form an intermediate double-stranded DNA during replication. A widespread but asymptomatic human circovirus called Transfusion Transmitted Virus (TTV) belongs to this group.

RNA viruses

Group III: Double-stranded RNA viruses

As with most RNA viruses, viruses in this group replicate in the cytoplasm, without having to use host cell polymerases as much as DNA viruses. This family has not been as well studied as the others and comprises two large subfamilies, the Reoviridae and the Birnaviridae. Replication is monocistronic and individual, in genome segments, which means that each gene encodes a single protein, unlike other viruses, whose transcription is more complex.

Groups IV and V: single-stranded RNA viruses

These viruses are made up of two types, both of which share the fact that replication takes place essentially in the cytoplasm, and that replication is not as dependent on the cell cycle as in other DNA viruses. This category of virus is one of the best studied, along with double-stranded DNA viruses.

Group IV: Polarity-positive single-stranded RNA viruses

(Virus (+) ssRNA or messenger RNA type)

The positive polarity of RNA viruses, and consequently the fact that all genes are defined as **sense,** means that the host's ribosomes can decrypt them directly and immediately synthesise proteins. These can be divided into two groups, both of which replicate in the cytoplasm:

• Viruses with polycistronic messenger RNA where genomic RNA forms mRNA and is translated into poly-proteins which are then cleaved to form mature proteins. This means that the gene can use several methods to produce proteins from the same strand of RNA, all with a view to reducing the size of its genome.

• Viruses with complex transcription, for which subgenomic mRNA, ribosome involvement and proteolytic processing of poly-proteins may be required. All these different mechanisms produce proteins from the same strand of RNA.

Examples in this category include the Astroviridae, Caliciviridae, Coronaviridae, Flaviviridae, Picornaviridae, Arteriviridae and Togaviridae families.

Group V: Negatively polarized single-stranded RNA viruses

Negative-polarity RNA viruses and even the set of genes defined as **antisense cannot be** directly decoded by the host polymerases to immediately produce proteins. Instead, they have to be transcribed by viral polymerases into a 'readable' positive polarity form. These can also be divided into two groups:

• Viruses containing a non-segmented genome for which the first stage of replication is the transcription of the (-) genome strand by the viral RNA-dependent polymerase into monocistronic mRNA encoding the various viral proteins. A (+) copy of the genome is then produced which serves as a template for the production of a (-) strand of the genome. Replication takes place in the cytoplasm.

• Viruses with a segmented genome for which replication occurs in the cell nucleus, and whose monocistronic RNA-dependent viral RNA polymerase produces an mRNA from each genome segment. The most important difference between the two is the location of replication.

Examples in this category include the families Arenaviridae, Orthomyxoviridae, Paramyxoviridae, Bunyaviridae, Filoviridae and Rhabdoviridae (this last family includes the rabies virus).

Reverse transcription DNA or RNA viruses:

These viruses use **reverse** *transcriptase* (**RT**), an enzyme that transcribes the genetic information of viruses from RNA into DNA, which can then be integrated into the host genome.

Group VI: Single-stranded RNA Retroviruses

A well-studied family in this class of viruses is that of the retroviruses. A key feature is the use of reverse transcriptase to convert positive-polarity RNA into DNA. Instead of using RNA for protein arrays, they use DNA to create arrays that are grafted into the host genome using integrase. Replication can then begin with the help of host cell polymerases. A well-studied example is HIV.

Group VII: Double-stranded DNA pararetroviruses

This small group of viruses, exemplified by the hepatitis B virus (part of the Hepadnaviridae family), comprises double-stranded viruses with a non-overlapping genome that is then completed to form a covalently linked tight ring (ccc DNA) that serves as a template for the production of viral mRNA and subgenomic RNA. The pregenomic RNA serves as a template for viral reverse transcriptase and genomic DNA production.

9.1.3 The concept of viral species

There are, of course, fundamental differences between viruses and other organisms, which explain the specific nomenclature adopted for the latter, in particular parasitism and the absence of sexual reproduction. The concept of a viral species is very special indeed!

The viral species is thus considered as a biological entity that forms a polythetical class of viruses, constituted by their descendants and delimited by the occupation of a particular ecological niche.

This means that the viral species is defined on the basis of a number of criteria, sometimes of a different order. However, not all the criteria need to be met in order to associate a given virus with the species.

This also means that a virus species can be defined on the basis of physical criteria, such as the nature of an RNA or the presence of a lipid envelope, but also biological and relational criteria, such as the host spectrum of a virus or the need for an insect vector.

Several criteria have been proposed to delimit a viral species:

- Example: see Phylogenetic tree of the Papillomaviridae family;
- physico-chemical properties of the virion ;
- the antigenic properties of viral proteins ;

- the "natural" host spectrum;
- cellular and tissue tropism ;
- pathogenesis and cytopathology ;
- the mode of transmission.

9.1.4 Specificity of viral taxonomy

There are, of course, fundamental differences between viruses and cellular organisms, which explain the specific nomenclature adopted for the latter. The names used to determine orders (suffix -virales, e.g. Mononegavirales), families (-viridae, Luteoviridae), subfamilies (- virinae, Paramyxovirinae) and genera (-virus, Retrovirus) are presented in Latin and italics, as for other organisms.

On the other hand, for the names of viral species, the official nomenclature is English-speaking and includes a vernacular designation of the virus as [disease] followed by [virus]. The viral species name is capitalized for the first term and for proper nouns included in the name, and is also written in italics (Tobacco mosaic virus, East African cassava mosaic virus).

The ICTV's eighth report confirms the existence of *3 orders, 73 families, 9 subfamilies, 287 genera and more than 5450 viruses grouped in 1950 species*. It also includes (by tradition) descriptions of other agents such as satellite viruses, viroids and agents of spongiform encephalopathies or prions.

New information technologies now make it possible to access a vast amount of information very quickly. The ICTV has developed a database that includes the names and descriptions of known viruses, as well as computer tools for identifying them.... Other websites are of interest for the value of the information they contain, such as the sequence bank sites (International Nucleotide Sequence Database Collaboration, which includes the GenBank and EMBL sites), the VIDE database for plant viruses and DPV.

10. Collections and conservation of viruses

As in the case of plants, it is impossible to keep a single type of virus. However, each viral species has a type (called a type species), which is a kind of reference chosen for that species. However, while it is impossible to conserve a virus in the form of a herbarium, several methods of conservation have been developed. There are several collections whose aim is to conserve and make available virus strains.

11. Viral cycle

The specific interaction between a viral protein and a cell receptor enables the virus to attach itself to the cell and introduce its genome. This plays two essential roles in the infected cell: firstly, it ensures the expression of viral proteins; secondly, it is replicated and then encapsidated to generate new infectious virions. These are released by the infected cell and can then spread the infection. Since replication of the viral genome and expression of viral proteins depend to a large extent on compartmentalised cellular machinery, the replication cycle of viruses in a cell varies greatly depending on the nature of the virus and its genome.

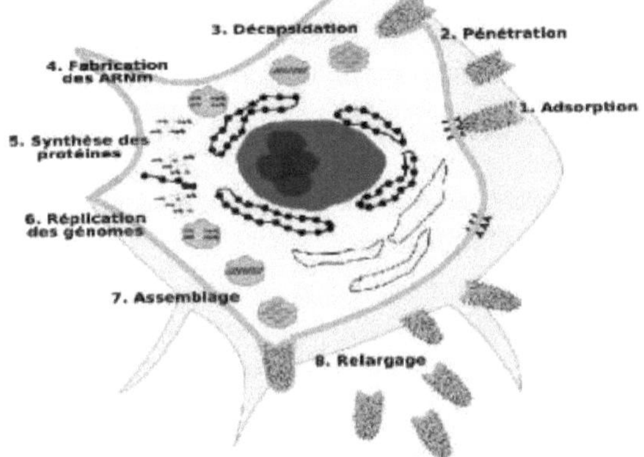

Figure 10: Development cycle of a virus in a host cell

1. **Attachment**, penetration and decapsidation, leading to internalisation of the viral genome in the target cell.
2. **Expression of the genes** and **replication** which, respectively, synthesise the proteins encoded by the viral genome and enable this genome to multiply.
3. **Assembly and release** that will lead to the production and release of infectious viral particles, capable of spreading the infection to other cells.

11.1. Virus attachment, penetration (entry) and decapsidation

The particular case of plant viruses is considered.

11.1.1 Attachment

- The first stage of infection is when the virus meets the target cell (adsorption). Attachment of the virus to the cell then occurs, following recognition of a receptor which is specific for the virus and usually corresponds to a surface protein on the target cell. Expression of this receptor is often restricted to certain types of cell or tissue. The receptor is therefore generally a crucial determinant of a virus's tropism.

- On the virus side, cell receptor recognition is performed by an external component of the virion. In the case of enveloped viruses, glycoproteins in the viral envelope are responsible for recognising a receptor on the cell to be infected. In the case of naked viruses, this interaction is mediated by capsid proteins.

Viral receptor and virus-receptor interaction

- The **receptor** used by viruses to bind to the host cell is a molecule on the surface of that cell. This may be a membrane protein (most often), a sugar (often itself attached to a protein), proteoglycans, a glycosaminoglycan such as herapan sulphate...

- This receptor is not expressed by the cell for the purpose of promoting infection by viruses. It generally plays an important physiological role for the cell in question: interaction with neighbouring cells, capture and transport of extracellular compounds .

the $CD4$ molecule, expressed by certain T lymphocytes and macrophages. This molecule is essential for the function of $CD4+$ T lymphocytes, enabling them to interact with the class II major histocompatibility complex molecules expressed by the antigen-presenting cells.

- The interaction between the virus and the receptor is specific. Each species of virus has evolved to recognise a particular receptor. There are, however, a few cases of viruses from different species using a common receptor.

- **Co-receptor**: Sometimes, infection of a cell involves recognition of more than one cell molecule. These are known as receptors and coreceptors.

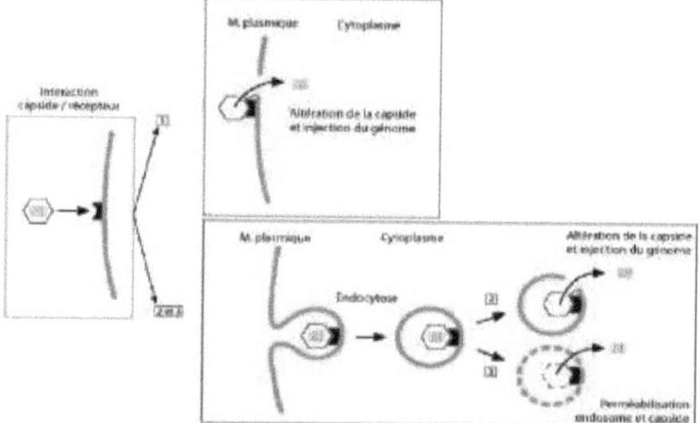

Figure 11: Attachment and penetration of a virus in a host cell

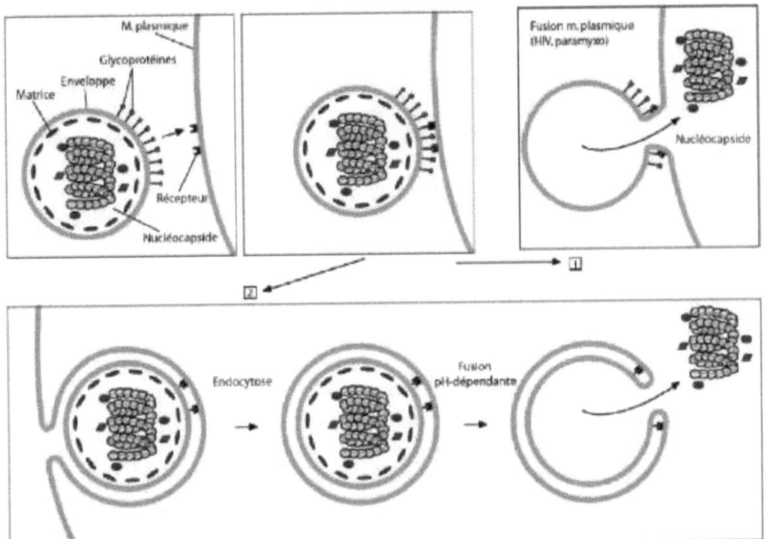

Figure 12: Penetration of a virus into a host cell and decapsidation

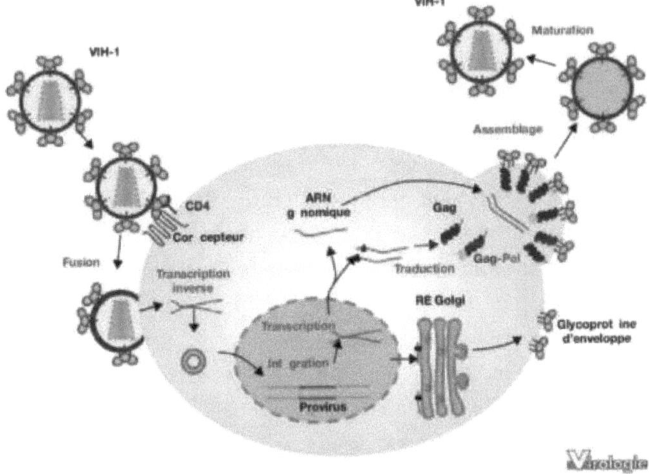

Figure 13: Penetration and replication of HIV in a host cell

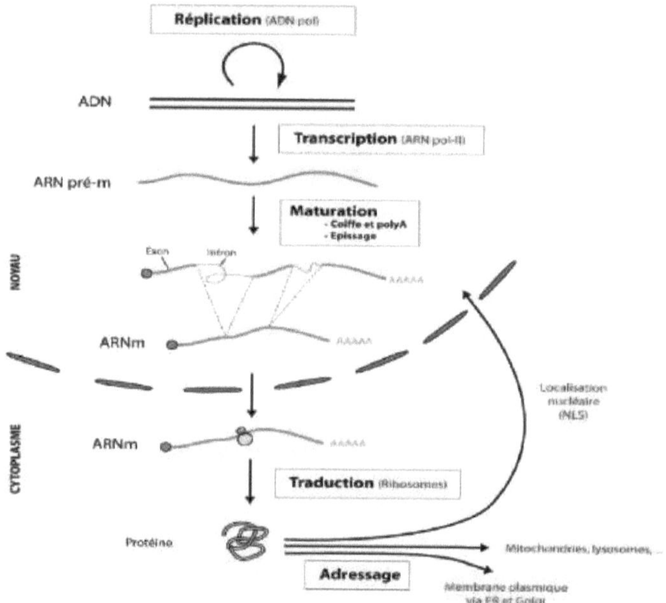

Figure 14: Replication, transcription and synthesis of viral proteins from a virus in a host cell

11.1.2 Virion - receptor interaction

The interaction between the virion and the receptor is a physical interaction that involves bonds of the same type as those that occur classically between biological molecules: formation of hydrogen bridges, electrostatic bonds (charge complementarity), interaction of hydrophobic domains, etc. The complementarity of the interacting domains is very precise and the affinity of the virion for the receptor can therefore be significant.

11.1.3 Penetration and decapsidation

The penetration and decapsidation stages result in the release of the viral genome into the target cell. As the viral genome penetrates the cell, it is partially or totally stripped of the proteins protecting it in the virion: this stripping process is called "decapsidation". The genome that ends up in the cytoplasm of the cell may be "free" (generally the case with RNA+ or DNA viruses), or remain associated with nucleoproteins in the form of a "nucleocapsid" (the case with RNA- viruses or, temporarily, retroviruses). Entry and decapsidation are dynamic phenomena, so they are difficult to study and relatively little is known about them. Depending on the nature of the virus, these stages vary considerably.

(Nucleocapsid diagrams. The nucleocapsid is the complex formed between the viral genome and the proteins that cover it.)

12. Naked viruses: naked (non-enveloped) viruses "inject" their genome into the cytoplasm of the cell. This may take place at the plasma membrane, following interaction of the capsid with the receptor. It can also take place after endocytosis. The genome is then "injected" through the wall of the endosome. It is thought that, in some cases, the capsid and endosome undergo alterations that make them permeable to the passage of the viral genome.

13. Enveloped viruses: what enveloped viruses have in common is that the entry of their genome into the cytoplasm of the host cell involves a fusion step between two membranes: the viral envelope and a host cell membrane. This fusion is carried out by certain glycoproteins in the virus envelope.

- For some of these viruses, binding to the receptor expressed on the cell surface directly fuses the viral envelope to the cell plasma membrane.

- For other enveloped viruses, interaction with the cell surface receptor induces endocytosis of the virus-receptor complex. Fusion then occurs between the viral envelope and the endosome membrane.

14. Terms and conditions for the entry of naked viruses (non-enveloped).

The proteins in the viral capsid interact with the receptor in the target cell. For some viruses (1), this triggers an "alteration" of the capsid, which "injects" the genome through the plasma membrane. For others (2 and 3), the alteration occurs after endocytosis of the virus-receptor complex. Alteration of the capsid leads either to injection of the virus genome through the endosome membrane (2), or to permeabilisation of the viral capsid and endosome (3) (Rhinoviruses seem to use the latter 2 strategies).

15. How enveloped viruses enter the cell.

The entry of enveloped viruses involves a fusion step between the virus envelope and a cell membrane. For enveloped viruses (1), such as the AIDS virus or paramyxoviruses, the fusion step occurs directly at the cell's plasma membrane. Other viruses (2) use the endocytosis pathway, via clathrin vesicles (coated pits), a type of endocytosis vesicle, or via the caveosome pathway. In general, the fall in pH of the endosomes causes a change in the conformation of the viral glycoprotein, which then triggers fusion of the virus membrane (envelope) and that of the endocytosis vesicle. Following fusion of the membranes, the nucleocapsid is released into the cytoplasm of the cell.

15.1 Membrane fusion

In both naked and enveloped viruses, two distinct signals and the involvement of several viral and cellular proteins are often required to ensure genome entry and decapsidation. The first signal is the interaction of the virus with the receptor. The second signal may be a drop in pH, temperature, or interaction with a secondary co-receptor required for viral entry. Some viruses bind sequentially, first to a receptor and then to a co-receptor.

Entry of the coxsackie virus into epithelial cells

Schematic diagram of coxsackie virus entry into epithelial cells

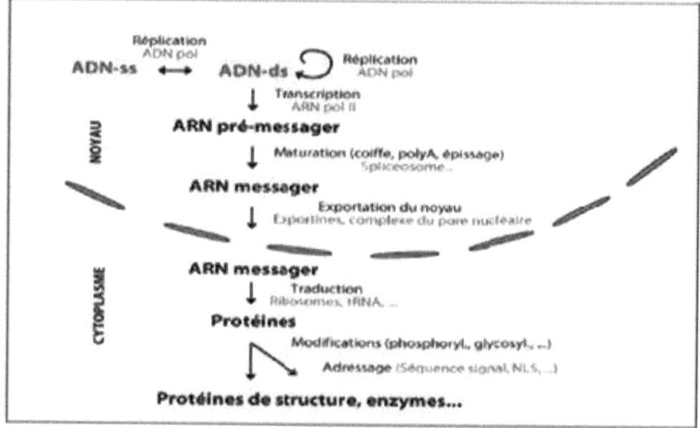

Figure 15: Replication, transcription and synthesis of coxsackie virus viral proteins in hdte epithelial cells

1. Attachment of the virus to the DAF receptor.
2. Activation of Abl and Fyn kinases.
3. Abl kinase induces migration of the virus-receptor complex.
4. Migration of the virus-receptor complex towards the side of the cell.
5. Interaction of the virus with the CAR receptor.
6. Fyn kinase induces endocytosis of the complex.
7. Endocytosis and decapsidation of the virus.

Panorama des virus d'intérêt médical et transmission

Virus à ADN		Virus à ARN	
Nus	**Enveloppés**	**Nus**	**Enveloppés**
Adéno D R ∞ Papilloma C S ∞ JC et BK virus ∞ Parvo B19 R	*Herpesviridae* ∞ : - Herpes simplex M S G ∞ - Varicelle-Zona R ∞ - CMV S M G T ∞ - E-BV S M G T ∞ - HHV-6 à 8 ∞ - Herpes B du singe C ***	Entérovirus D HAV D Rhinovirus R Rotavirus D Astrovirus D Calicivirus D	Myxovirus Influenza : - Grippe R *Paramyxoviridae* : - Para-influenza R - Oreillons R - Rougeole R (∞) - RS R *Coronaviridae* R Rubéole R *Flaviviridae* : - Fièvre jaune C *** - HCV G T C ∞ Rage C (G) *** Lassa, Hanta R *** Ebola, Marbourg R *** *Retroviridae* ∞ : - HIV-1 et2 S M G T C ∞ *** - HTLV-1 et2 S M G T C ∞
Virus complexes : HBV S M T C ∞ *Poxviridae* : -Variole R *** -Vaccine C			

D Voie digestive
R Voie respiratoire
M Echanges mère-enfant
S Voie sexuelle
T Transfusion sanguine
G Greffe
C Voie transcutanée
∞ Infection virale persistante

A part, les prions ou ATNC *** (protéine autoréplicable ?) G et sans doute D, T pour l'ESB

*** Haute mortalité

Cours I – illustration 3/19

16. Special case of plant viruses

Plants are distinguished from most other organisms by their thick cell walls, which are made up of cellulose and other materials. There is no endocytosis, and membrane fusion processes are not common. As a result, the entry of plant viruses most often occurs by breaking and entering: either via insects, nematodes or protozoa that parasitise plants, or by mechanical inoculation. Some plant viruses are transmitted efficiently by vegetative means (cuttings, tubers, etc.) or via pollen and seed.

The way plant viruses propagate within the organism also varies significantly from that of animal viruses or bacteriophages. Plant cells are connected by plasmodesmata, pores whose diameter (close to 20 nm) is variable. The virus can therefore propagate from one cell to another without passing through an "extracellular virion" stage. It thus bypasses the release and entry stages. To transmit their genome or nucleocapsid from one cell to another, plant viruses generally express one or more movement proteins that modulate the functioning of plasmodesmata. In addition, they can implement complex mechanisms that favour their transmission from plant to plant by their vector.

16.1 Viral gene expression and replication

-   Within the cell, the viral genome plays two distinct roles. Firstly, it is used to ensure the expression of viral proteins, which are necessary for replication of the virus and then for the formation of new viral particles. Secondly, it is multiplied ("replicated") before being encapsidated to form new viral particles.

-   the nature of the viral genome largely determines the strategy that will be followed by each virus to make the best use of the cellular machinery to ensure viral gene expression and genome replication. It should be noted that the cell is a compartmental space in which different stages of replication, gene expression or protein addressing can occur in distinct compartments.

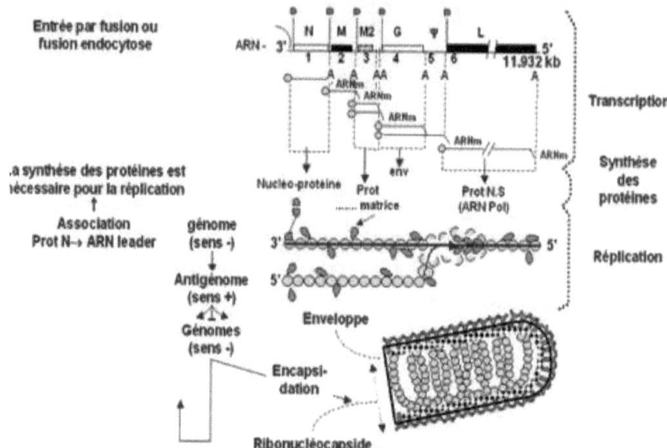

Figure 16: Expression of viral genes and replication of a virus in an hdte cell

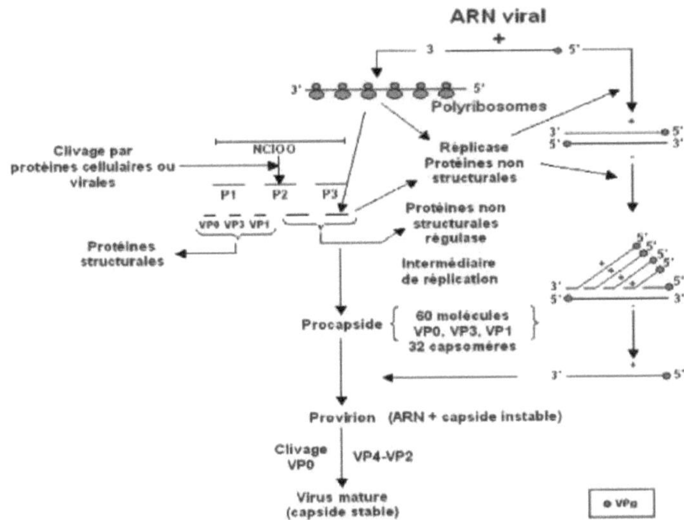

Figure 17: Expression of viral genes and replication of an RNA+ virus in a hdte cell

16. Compartmentalisation of the eukaryotic cell

In a eukaryotic cell, the replication, transcription and maturation of messenger RNAs (mRNAs) take place in the nucleus. Translation of the mRNAs into proteins takes place in the cytoplasm. The proteins produced are then sent to the compartment where they are to exert their action: nucleus, cytoplasm, endoplasmic reticulum, lysosomes, mitochondria, membrane, etc.

17.1 DNA viruses

Double-stranded DNA viruses (group I according to Baltimore) generally use the cellular machinery, both for their replication and for the transcription of their genes into mRNA and then for the maturation of these mRNAs. Their cycle is therefore nuclear.

Despite their nuclear cycle (and therefore despite the presence of the cellular enzymes required for replication and transcription), herpes viruses code for their own DNA polymerase, giving them a degree of independence from the cell cycle for replication. Viruses in the Poxviridae family, such as vaccinia virus, are a notable exception. They have a cytoplasmic replication cycle. These viruses therefore code for all the enzymes responsible for replication of viral DNA and for the enzymes needed to produce mRNA.

In the case of DNA viruses, gene expression is regulated over time. Some viruses, such as parvoviruses, have a single-stranded genome (single-stranded DNA) (Baltimore group II). However, these viruses use cellular DNA polymerases for replication (which transiently takes a double-stranded form), and cellular RNA polymerase II for transcription of the genome into mRNA.

- the dependence of viral replication on cellular enzymes means that the cells themselves must be in the replication phase. For example, parvoviruses preferentially, if not exclusively, infect cells in mitosis.

17.1.1 Replication of DNA viruses.

The genome of DNA viruses (with the exception of Poxviruses) is replicated in the nucleus. It is generally replicated and transcribed by cellular polymerases (in blue on the diagram). The genome of single-stranded DNA viruses (ssDNA) is converted to double-stranded DNA (dsDNA) by the cellular polymerase. The cellular machinery is responsible for mRNA maturation and translation.

17.2. RNA viruses

RNA viruses have a cytoplasmic replication cycle. No nuclear enzyme in the cell can be useful for replication or transcription, since to date no RNA-dependent RNA polymerase capable of transcribing long segments of RNA has been found in mammalian cells. RNA viruses therefore code for their own polymerase. Viral polymerase is generally a multifunctional enzyme that replicates the genome, transcribes into mRNA and sometimes adds caps and polya tails to messenger RNAs. By replicating in the cytoplasm of cells, they can exploit the presence of cellular ribosomes to ensure the translation of their mRNAs. Some RNA viruses, such as influenza virus (Orthomyxovirus) or Borna virus (family Bornaviridae, related to Rhabdoviridae), make an exception to this rule by replicating in the cell nucleus. These viruses exploit the cell's splicing machinery (nucleus) to increase their coding capacity, by means of differential splicing.

RNA-polymerase / RNA-dependent activities

17.2.1. These polymerases synthesise RNA molecules that are complementary to an RNA template. They can synthesise strands of RNA+ or RNA- corresponding to genomes or "antigenomes", copying the entire template. The recognised initiation signal is then a "promoter" located at the end of the template molecule.

17.2.2. These polymerases can also recognise promoters and transcription termination signals internal to the template molecule, in particular to synthetise subgenomic mRNAs. Some of these enzymes are capable of forming a cap at the 5' end and a poly A tail at the 3' end of transcribed mRNA molecules.

17.2.3.   **RNA-positive viruses (RNA+) (Baltimore group IV)**

RNAs that have the same polarity as the messenger RNA that codes for proteins are known as "positive polarity RNA" or "RNA+". Most (if not all) RNA+ viruses have a genome that possesses the signals required to be translated directly by the host cell's ribosomes.

In some viruses, such as Picornaviruses or Flaviviruses, all viral proteins can be synthesised from genomic RNA. In these viruses, a single open reading frame (OrF) synthesises a polyprotein which is cleaved by an auto-catalytic process (viral proteases contained in the polyprotein) to provide all the viral proteins.

In other viruses, such as Togaviruses or Coronaviruses, only some of the proteins can be produced using the genome as mRNA. These proteins include viral polymerase. This polymerase synthesises a complementary strand to the genome (antigenome) and then synthesises genomic or subgenomic mRNA, which is used to translate the other viral proteins (often the structural proteins).

17.2.4.   Replication of RNA+ viruses (without sub-genomic mRNA)

The genome of some RNA+ viruses codes for a single polyprotein which is cleaved by one or more viral protease(s) to give all the proteins required for the viral cycle. In this case, all the virus proteins are translated from the genomic RNA. The genome is replicated by a viral polymerase, which copies the genomic RNA (+) into anti-genomic RNA (-) and then copies this anti-genomic RNA back into genomic RNA.

The replication cycle of RNA+ viruses is cytoplasmic. Replication and transcription are carried out by a viral polymerase. Translation is carried out by the cellular machinery.

Because the genome of RNA+ viruses is generally recognised as mRNA by the cellular machinery, it is easy to obtain infectious clones of these viruses.

17.2.5.   Negative RNA (RNA-) viruses (Baltimore Group V)

- Negative RNA viruses (RNA⁻) (Rhabdovirus, Paramyxovirus, Orthomyxovirus) have a genome whose polarity is complementary to that of messenger RNAs. As in the case of RNA+ viruses, the polymerase used by these viruses is encoded by the viral genome. It is an RNA-dependent RNA polymerase which carries out the functions of transcription and genome replication.

- For transcription, the polymerase encoded by the virus forms, from the RNA genome⁻, subgenomic mRNAs corresponding to each "gene". These mRNAs are then taken over by the cellular ribosomes for translation. For replication, the same polymerase synthesises an anti-genome (a complementary copy of the entire genome) which then serves as a template for the production of new RNA genomes⁻.

- 7.2.2.1 Replication of RNA viruses (-)

-The genome of RNA- viruses is transcribed into subgenomic mRNA by the viral RNA polymerase. This same RNA polymerase also replicates the viral genome (RNA- > RNA+ > RNA-). Translation of the mRNAs is carried out by cellular ribosomes.

Some viruses are called ambisense RNA viruses (as in the case of certain arboviruses such as Bunyaviruses, Tospoviruses or Arenaviruses) because their genome cannot be attributed any polarity. In fact, their genome is similar to that of certain RNA viruses⁻. However, while the majority of messenger RNAs are complementary to the genome, some proteins are produced by mRNAs that have the same polarities as the genome. Strictly speaking, therefore, it is not possible to define a + strand and a - strand, hence the name "ambisense RNA virus".

Note that a transcription step is required to produce viral mRNA from the RNA genome⁻. Viral RNA polymerase cannot therefore be produced by infected cells without prior transcription of the genome into mRNA (by viral RNA polymerase). RNA viruses⁻ are therefore obliged to carry a few copies of the viral polymerase in their virion in order to initiate their replication cycle in the host cell.

17.2.6.　Double-stranded RNA viruses (dsRNA) (Baltimore Group III)

Some viruses, such as Reoviruses, Rotaviruses and Birnaviruses, have a segmented genome made up of double-stranded RNA. The RNA- strand serves as a template for the production of the various messenger RNAs. As with other RNA viruses, a polymerase encoded by the virus is responsible for transcription and replication of the genome.

17.3. Viruses using reverse transcriptase

Retroviruses (HIv, Htlv, etc.), Hepadnaviruses (hepatitis B virus) and Caulimoviruses (cauliflower mosaic virus) are unique in that they code for a reverse transcriptase (RT) which, during their replication cycle, converts viral RNA+ into double-stranded DNA (retrotranscription). In the case of retroviruses, it is the RNA molecule that is encapsidated to form the virion, and retrotranscription takes place when the viral nucleocapsid enters the cytoplasm of the infected cell (Baltimore group VI). In the other two cases (Hepadnavirus and Caulimovirus), retrotranscription occurs when the virus leaves the cytoplasm of the infected cell. It is therefore a DNA genome that is encapsidated to form virions (group VII according to Baltimore).

17.3.1 Replication of retroviruses

The retrovirus genome is an mRNA initially transcribed by cellular RNA polymerase II. Reverse transcriptase recopies this RNA into double-stranded DNA, which migrates into the nucleus and is integrated into the cell's genome. The viral mRNA transcribed by cellular RNA polymerase II can either undergo splicing or remain unspliced, and is exported to the cytoplasm where it is translated by cellular ribosomes.

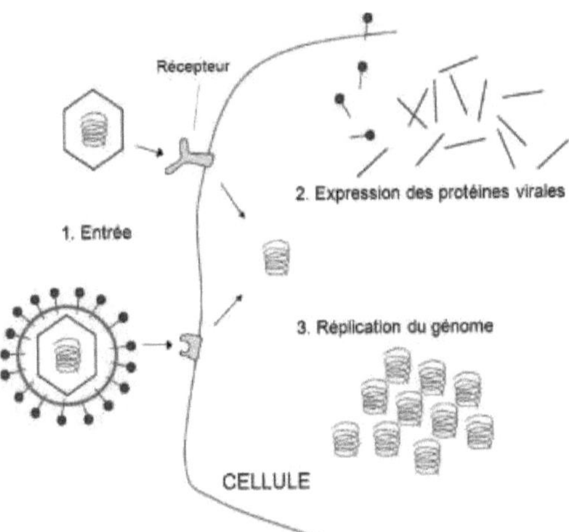

Figure 18: Viral receptors, protein expression and genome replication of an envelope virus in a hdte cell

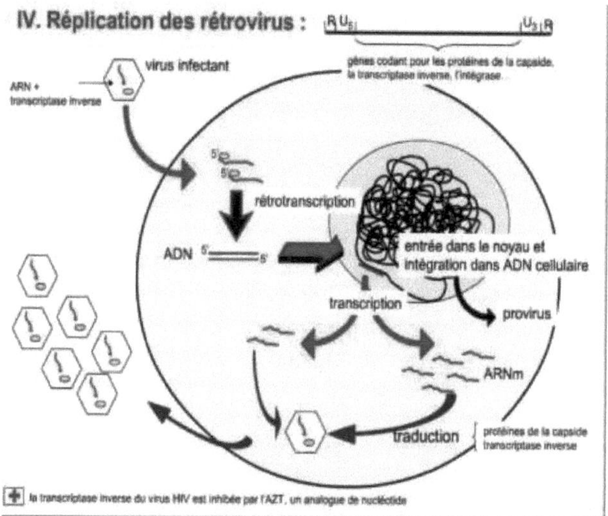

Figure 19: Penetration, Replication, Transcription, Protein synthesis and Virus exit from the hdte cell

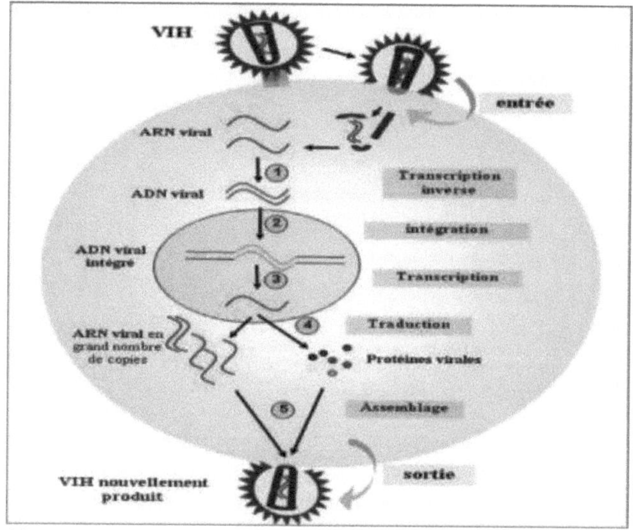

Figure 20: Penetration, Replication, Transcription, Protein Synthesis and Virus Exit from an hdte cell

17.4   Assembly and release of new viruses

After replication of the viral genome and synthesis of the structural proteins, the virions are assembled and released by the host cell, so that they can spread to other cells or organisms.

17.5 Non-enveloped (naked) viruses

It appears that the assembly of non-enveloped viruses is the result of a highly efficient self-assembly process combining the viral genome and capsid proteins. Mature viruses accumulate in the nucleus or cytoplasm of the infected cell and are then released by cell lysis. This lysis occurs as a result of the structural and metabolic disorganisation of the infected cell caused by the massive production of viral elements at the expense of cellular proteins. At present, there is no clear evidence that naked viruses are released in the absence of cell lysis. However, this type of event is sometimes suspected.

17.6 Virus envelopes

Enveloped viruses are released from infected cells by budding. The glycoproteins encoded by these viruses are inserted into the cell plasma membrane.

The nucleocapsids assembled in the nucleus or cytoplasm interact with the glycoprotein-coated regions of the membrane. This interaction is usually mediated by a matrix protein located on the inner surface of the plasma membrane.

Budding then begins, leading to the release of nucleocapsids surrounded by an envelope corresponding to the plasma membrane of the producing cell, into which the viral glycoproteins are inserted.

Unlike release by cell lysis, the production of enveloped viruses by budding is not necessarily accompanied by the death of the producing cell.

17.7 Plant viruses

Plant viruses are again an exception. They depend on their vector or on mechanical intrusion (agricultural machinery, etc.) to ensure that they leave the plant cell, which is surrounded by a wall.

17.8 Viral and infectious particles

Not all virions produced by infected cells are functional. In fact, as a result of polymerase errors, it is possible that the capsid genome in the virion is not functional. It is also possible that the capsid has not been formed correctly or that any other accidental alteration renders the virus non-infectious. We therefore distinguish between the terms "viral particle", which corresponds to a complete but not necessarily functional virion, and "infectious particle", which corresponds to a virion that is actually capable of infecting another cell. The ratio of infectious particle to viral particle is sometimes very low (often 1/100 or even 1/1000).

17.8.1 Genetic variations

When we consider a population of viruses, we are generally dealing with a group of viruses with a degree of heterogeneity. We are dealing with billions of individuals that may have evolved in different ways. Not all of them will be genetically identical, even within the same host. The main mechanisms that lead to genetic changes in viruses are mutation, recombination and a variant of this, reassortment. Furthermore, in the course of their evolution, viruses can lose certain genetic elements (deletion) or, on the contrary, acquire them, for example from a cell (insertion). Viruses also evolve through the inversion and repetition of certain genomic fragments.

17.8.2 Changes

In any nucleic acid chain, errors can occur during transcription, leading to a change in the genetic information (mutation). This phenomenon therefore only occurs in replicating viruses.

In general, these mutations occur more frequently in RNA viruses. The polymerases they use (RNA-dependent polymerases) have no corrective mechanisms (3'-5' exonuclease activity), unlike DNA-dependent polymerases, which must ensure the continuity of the genetic code.

If a mutation is neutral or even favourable, it can persist. When one nucleotide is substituted by another, this is known as a point mutation. This can be silent when it does not lead to changes in the amino acid chain and does not affect a sequence or structure of the nucleic acids important for replication.

When a mutation in a coding region results in a change of amino acid, this may be deleterious, confer an advantage on the virus or have no effect at all. If the mutation favours the virus because it is better adapted to external circumstances, the mutant will gradually take over from the original virus. The external factors that lead to selection may be the immune system, the change in environment brought about by a new host or antiviral drugs. When a varied population of different mutant viruses develops within a single host, this is known as a "quasi-species".

Cloning can be used to distinguish between the different mutants. If we carry out global nucleotide sequencing, we obtain a consensus sequence, made up of a mixture of the different dominant sequences in the viral population. The quasi-species phenomenon enables the virus to adapt rapidly to changing circumstances. This is how, in infection with the AIDS virus (HIV) or the hepatitis C virus, the virus manages to continually evade the immune system and establish a chronic infection. The existence of quasi-species in the case of HIV means that the virus is highly plastic, rapidly evading treatment when it fails.

In terms of the world's population, we can see that these progressive evolutions result in sets of viruses that can be genetically differentiated in the form of different genotypes, possibly grouped into genogroups.

One particular phenomenon is that observed with the human influenza virus.

Under the immune pressure of the world's human population, the virus evolves by gradually accumulating mutations, resulting in antigenic slippage or "drift". This will enable it to periodically find a human population that is not immune to the mutant virus present. This is why vaccines against the flu virus need to be adapted every year.

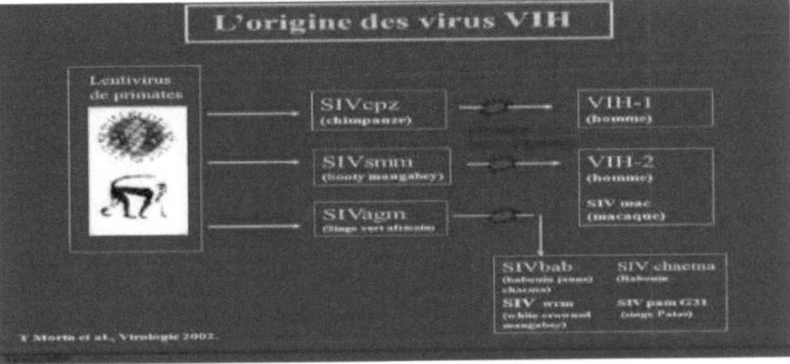

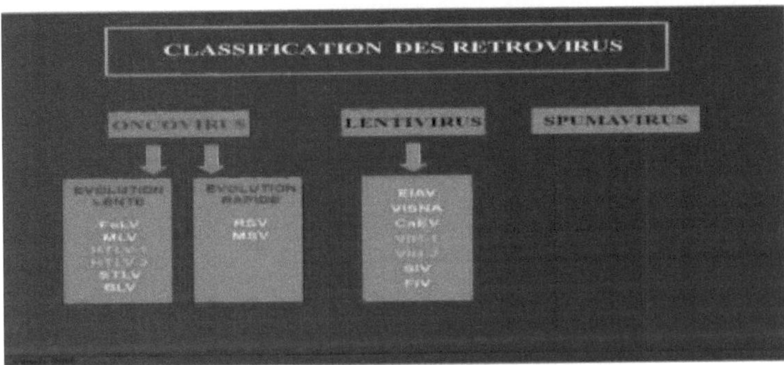

Figure 21: Origin of HIV and classification of Retroviruses

17.8.3 Mutation by substitution

A substitution mutation is the replacement of one nucleotide by another. These mutations can be 1) transitions, where a purine replaces a purine (a or g) or a pyrimidine replaces a pyrimidine (C or t/u), or 2) transversions, where the exchange occurs between purine and pyrimidine. The latter occur less frequently.

17.8.3.1 Amino acid: When a mutation occurs in a coding part of the genome, there is not always a corresponding change of amino acid. The genetic code gives combinations of three nucleotides (= codon) coding for an amino acid (see table). As there are 4 different nucleotides, there are 64 possible combinations, but there are only 20 natural amino acids. For the same amino acid there are different codons possible (up to 6). A change of the third nucleotide in a triplet is unlikely to result in a change of amino acid, whereas changes of the first or

41

second nucleotide almost always result in a change.

17.8.3.2 Viral sequencing of a mixture

When we are dealing with a quasi-species, we are in the presence of a cloud of mutants, i.e. viruses with different genetic codes. Global sequencing (the definition of the nucleotide sequence) will give us a false idea of the mutants present, or will not allow us to draw any conclusions. Either the different variants are present in too small quantities (less than 20%) and we will not detect them, or there will be confusion over the codons. Suppose that a global sequencing tells us that there is a mixture, for example: T, T or G, T or G. The possible codons are TTT (coding for phenylalanine), TGG (tryptophan), TTG (leucine), TGT (cysteine), without us knowing which is actually present. To clearly define the population, the different mutants will be analysed separately by cloning them (by introducing viral genes into bacteria, using phages or plasmids, we can separate the genomes by separating the bacteria that contain them).

17.8.3 .3 Recombinations and reassortments

Similar viruses can exchange homologous (and sometimes non-homologous) genetic regions when infecting the same cell. The polymerase will switch from one strand to the other during replication. This can be facilitated, for example, by the simultaneous presence of two copies of the genome, as seen in HIV (diploid virus). This phenomenon allows the virus to change rapidly. Over the course of its evolution, HIV has given rise to numerous subtypes. In human populations where there is high transmission of this virus, mixed infections with several subtypes are not uncommon, and recombinant forms with characteristics of both "parent" viruses have emerged.

One particular form is reassortment, which can occur when the virus genome is segmented, as is the case with the influenza virus. The influenza A virus is originally a waterfowl virus, and in these species many different types of this virus are present. In rare cases, mixed infections can occur involving, for example, a bird virus and a human virus. When this happens, the segments of the viruses, of which there are eight, can mix and give rise to numerous variants. The viruses resulting from these mixed infections have various combinations of segments from the two original viruses (see figure) (theoretically 28 = 256 possible combinations). It is possible that one of these new viruses is particularly well adapted to the human host and gives rise to what is known as an influenza pandemic, i.e. a rapid expansion of the virus in the human population that is not immune to this new virus. Reassortments with less significant consequences also exist: an influenza A H1N2 virus (hemagglutinin 1, neuraminidase 2) has recently been described, which is a virus reassorted from the two circulating types of influenza A, H1N1 and H3N2.

SPECIAL VIROLOGY

1. HIV/AIDS

1.1 **HIV :**

1.2 **Definition**: Human immunodeficiency viruses (HIV) are viruses belonging to the retrovirus family and the lentivirus subfamily. As with all retroviruses, the HIV genome contains three genes encoding different viral proteins: gag, pol and env, coding respectively for structural proteins, viral enzymes and envelope glycoproteins.

There are two viral types of HIV, HIV-1 and HIV-2, the result of two different zoonotic transmissions. HIV-1 is widespread throughout the world. It is the cause of the AIDS pandemic and poses a major public health problem on every continent.

HIV-2 has a much more limited distribution and HIV-2 infection has not developed into an epidemic. The aim of this course is to describe the characteristics of HIV and its replication in order to better understand the pathogenesis of this viral infection and the interactions between the virus and its host:

- The first phase of the cycle begins with the attachment of the virus to the CD4 molecule and ends with the integration of the proviral deoxyribonucleic acid (DNA) into the cell genome;

- the second phase of the cycle begins with the transcription of double-stranded viral DNA and ends with the emergence of new virions by budding on the cell surface.

HIV-1 has a tropism for cells carrying the CD4 molecule, the main ones being CD4 lymphocytes, monocytes and macrophages... The genetic variability of HIV is extreme.

HIV-1, isolated in 1983, comprises four groups: M, N, O and P. Group M comprises nine subtypes (A to K) and multiple CRFs (recombinant circulating forms), while HIV-2 comprises nine groups (A to I).

Within HIV-1, M is the major group, the other three groups are in the minority, mainly identified in Central Africa, particularly in Cameroon, called O (Outler) and N (non M, non 0) and more recently group P. Each of these groups corresponds to a distinct transmission event in monkey SIV viruses. The chimpanzee simian virus, SIVcpz, and HIV-1 are 80-90% homologous. To date, >95 CRFs, resulting from recombination between two or more HIV-1 group M subtypes, have been described.

HIV-2, isolated in 1986, is closely related to the simian sooty mangabey virus (SIVsmm). It is classified into nine distinct groups based on the homology of the gag and env gene sequences. HIV-2 is naturally resistant to natural reverse transcriptase (RT) inhibitors and fusion inhibitors. It is mainly present in West Africa, in Mozambique and Angola, Portuguese excolonies, from Guinea-Bissau onwards, and outside Africa, in India and Brazil.

1.1.2 Structure of HIV: HIV is a retrovirus of the lentivirus genus (from the Latin *lentus* meaning "slow"), characterised by a long incubation period and consequently a slow progression of the disease.

HIV-1 is a spherical virus with an average diameter of 145 nanometres. Like many viruses that infect animals, it has an envelope made up of a fragment of the membrane of the infected cell. Trimeric envelope glycoproteins (Env) are inserted into this lipid envelope. Each Env protein consists of 2 subunits: a surface subunit gp120 and a transmembrane subunit gp41. The surface of an HIV virus is thought to contain an average of just 14 Env trimeres. When the virus attaches itself to the cell, the gp120 Env protein binds to a CD4 receptor present on the surface of the immune system's CD4+ cells. For this reason, HIV only infects cells that have this receptor on their surface, the vast majority of which are CD4+ lymphocytes.

Inside the envelope is a protein matrix (MA) made up of p17 proteins and, still inside, the capsid (CA) made up of p24 proteins. It is this last type of protein, along with gp41 and gp120, that is used in HIV western blot tests. The p7 nucleocapsid (NC) proteins protect viral RNA by coating it. The p6 protein is excluded from the capsid and lies between the matrix and the capsid, allowing newly formed viruses to bud out of the cell.

The HIV genome, contained in the capsid, consists of a single strand of RNA in duplicate (9181 nucleotides)[42] , accompanied by enzymes:

•   p66/p51 reverse transcriptase or retrotranscriptase, which retrotranscribes viral RNA into viral DNA;

•   p32 integrase, which integrates viral DNA into cellular DNA;

•   protease p12, which participates in virus assembly by cleaving the protein precursors Gag p55 and Gag-Pol p160. The protease is present in the capsid[43] .

These three enzymes are the main targets of antiretroviral treatments, because they are specific to retroviruses.

The HIV genome is made up of nine genes. The three main genes are *gag, pol* and *env,* which define the structure of the virus and are common to all retroviruses. The other six genes are *tat, rev, nef, vif, vpr* and *vpu* (or *vpx* for HIV-2), which code for regulatory proteins.

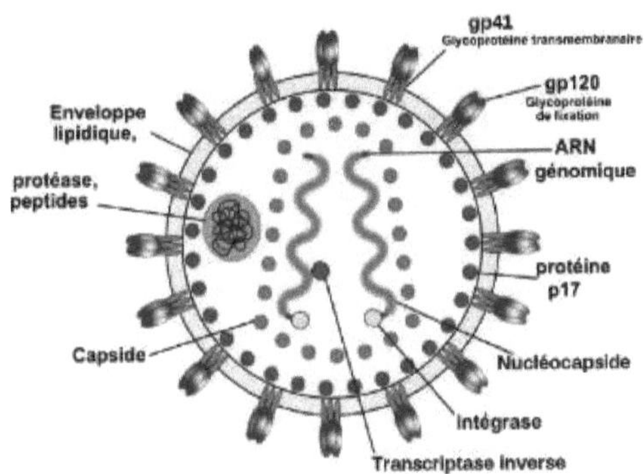

Figure 22: Structure of the AIDS virus.

1.1.3 Systematic position :

Type: HIV-1 and HIV-2

Group : VI

Family: Retroviridea

Subfamily: Orthoretrovirinea

Genre: Lentivirus

Species: human immunodeficiency virus.

1.1.4 Background: The AIDS epidemic began on 5 June 1981, when the US CDC announced an upsurge in cases of *Pneumocystis carinii* pneumonia and Kaposi's sarcoma in the cities of Los Angeles, San Francisco and New York. These two diseases are particularly common in immunocompromised people. It is precisely in these patients that the level of T4 lymphocytes plummets. These cells play an essential role in the immune system. The first patients were all homosexuals, so this syndrome, which was not yet called AIDS, was provisionally called the *gay syndrome* or *gay cancer*. One of the first suggested causes of this immunodepression was poppers, a vasodilator widely used by homosexuals. But in the months that followed, other people were infected, including injection drug users, haemophiliacs and Haitians. This discovery showed that poppers was not the cause, and an infectious origin was increasingly accepted. All that remains is to find the infectious agent.

1.1.5 Transmission: HIV is present in many body fluids. It has been found in saliva, tears and urine, but in concentrations insufficient to record cases of transmission. Transmission via these fluids is therefore considered negligible. On the other hand, quantities of HIV large enough to trigger infection have been detected in blood, breast milk, cyprine, sperm, as well as the fluid preceding ejaculation, and the concentration of the virus in genital secretions (sperm and cervical secretions in women) are good predictors of the risk of HIV transmission to another person.

Consequently, the three modes of contamination are :

• unprotected sex. Whether heterosexual or homosexual, unprotected sex accounts for the largest proportion of infections;

• contact with contaminated material at :

o drug addicts, by injection,

o tattoos, due to poor hygiene of the equipment,

o transfusions,

o healthcare staff ;

- mother-to-child transmission, during pregnancy, childbirth and breastfeeding. Without treatment and with natural childbirth, the rate of transmission varies between 10% and 40%, depending on the study. It is during childbirth that the risk of infection is highest (65% of all cases of infection)[55] . Treatment and a possible cesarean section can reduce this figure to 1%.

1.1.6 Multiplication

HIV infects **cells** that carry the CD4 antigen on their surface (T **cells**, but also monocytes and certain other **cells**). A high-affinity interaction occurs between the gp120 protein of the viral envelope and the CD4 membrane marker of the **target cell**.

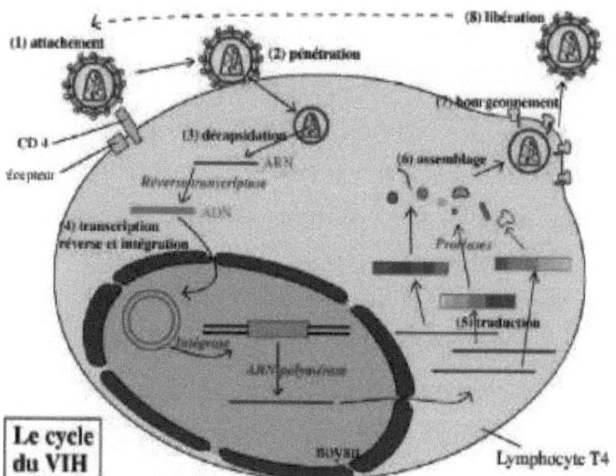

Figure 23: HIV development cycle in a hdte cell

1.2 AIDS

1.2.1 Definition: AIDS is acquired immunodeficiency syndrome. **HIV** infection affects the immune system, i.e. the body's natural defences against disease.

1.2.3 Geographical distribution :

Despite remarkable progress in the global fight against HIV since 2000, new infections and AIDS-related deaths remain at high levels. AIDS remains one of the deadliest pandemics of our time: 690,000 people died of AIDS-related illnesses in 2020, the same figure as in 2019. The total number of new infections has fallen by only 31% since 2010, far short of the 75% target set by the United Nations General Assembly in 2016.

In 2020, 84% of all people living with HIV knew their serological status, 73% had access to treatment and 66% had an independent viral load.

This is a good result, even if the 90-9090 screening and treatment targets have not been achieved in all countries.

The new "Global AIDS Strategy 2021-2026: Ending Inequality, Eradicating AIDS", is on track to eliminate AIDS by 2030 although, according to UNAIDS, inequality within countries is preventing the world from ending AIDS by 2030.

The resources available in low- and middle-income countries were of the same order as in 2019, less than US$ 19 billion, whereas UNAIDS estimated that US$ 26.2 billion were needed for the AIDS response in 2020. Under-investment in low- and middle-income countries is the main reason why the 2020 targets have not been met. The target for 2025 is US$29 billion.

The global HIV situation is as follows:

Regions	PLWHA	New infections	Deaths liesau HIV	People on ART	PLHIV with access to ART (%)
East and Southern Africa	0,6*	670 000	310 000	15,0*	77 %
West and Central Africa	4,7*	200 000	150 000	3,5*	73 %
Asiaand Pacific	5,7*	280 000	140 000	3,6*	64 %
Latin America	2,1*	110 000	32 000	1,4*	65 %
Caribbean	330 000	13 000	6 000	220 000	67 %
Medium-Orient +	230 000	16 000	7900	93 000	41 %

*in millions

2/3 (67%) of PLHIV live in sub-Saharan Africa. Eastern and Southern Africa remains the region hardest hit by HIV. It accounts for around 55% of all people and two-thirds of all children living with HIV. It is also the region that has made the greatest progress against the HIV epidemic since 2010. New infections fell by 43% between 2010 and 2020, and by 64% among children aged 0-14, the biggest reductions of any region. In West and Central Africa, the response to HIV is improving, but not fast enough to end AIDS as a threat to public health by 2030. While the number of new infections in 2020 has fallen by 37% compared with 2010, the region accounts for more than a third of new HIV infections among children worldwide, reflecting gaps in efforts to prevent transmission.

1.2.3. Transmission modes

HIV can be transmitted by close, unprotected contact with the body fluids of an infected person: blood, breast milk, sperm and vaginal secretions. Several modes of transmission have been described:

1.2.3.1 Heterosexual transmission

It is predominant in tropical regions, particularly in sub-Saharan Africa. Six out of seven new infections in adolescents aged 16 to 19 are among girls. Young women aged 15 to 24 are twice as likely to be living with HIV as men. More than a third (35%) of women worldwide have experienced physical and/or sexual violence by an intimate partner or sexual violence by a non-partner at some point in their lives. In some regions, women who have experienced physical or sexual violence by an intimate partner are 1.5 times more likely to contract HIV than women who have not experienced violence. Women and girls will account for around 50% of all new HIV infections worldwide in 2020.

In sub-Saharan Africa, women and girls accounted for 63% of all new HIV infections.

1.2.3.2 Transmission from mother to child

Infection among young women explains the frequency of mother-to-child transmission (MTCT) in developing countries. New infections among children fell by 54% between 2010 and 2020, mainly as a result of increased provision of ARVs to pregnant and breastfeeding women. It is estimated that 85% of pregnant women living with HIV worldwide will have received ART by 2020 to prevent mother-to-child vertical transmission and to preserve their own lives. However, there are still significant disparities within many countries in West and Central Africa: 24% of pregnant women not receiving ART are in just one country, Nigeria, and a further 33% live elsewhere in West and Central Africa.

Gaps in the screening of infants and children exposed to HIV mean that more than two-fifths of children living with HIV remain undiagnosed. Almost 800,000 children (aged 0-14) living with HIV were not tested in 2020. Sub-Saharan Africa accounts for 89% of new HIV infections in children and 88% of children and adolescents living with HIV worldwide.

1.2.3.3 Transmission among key populations

The importance of key populations in the transmission of HIV was reported according to region.

Homosexuals are 25 times more likely to contract HIV than heterosexuals; sex workers are 26 times more likely to contract HIV than women in the general population; transsexual women are 34 times more likely than other adults; people who inject drugs are 35 times more likely to contract HIV than people who do not inject drugs.

Overall, key populations and their sexual partners account for 65% of HIV infections worldwide in 2020 and 93% of infections outside sub-Saharan Africa. These key populations are marginalised. People in prison often do not benefit from any HIV care services.

1.2.4 Diagnosis of HIV/AIDS infection

Forty years after the start of the HIV/AIDS epidemic, the issue of screening remains crucial, with around 20% of seropositive people worldwide still unaware of their serological status. The situation has improved in recent years with the advent of RDTs and, more recently, the development of self-tests.

1.2.4.1 The screening and biological diagnosis of HIV infection in developed countries is traditionally based on a two-stage strategy: screening analysis, followed by confirmatory analysis. The screening test uses the combined ELISA, which simultaneously detects anti-HIV1 and anti-HIV2 antibodies and the HIV-1 p24 antigen (with a minimum detection threshold for the p24 antigen of 2 IU/mL).

A negative test leads to the conclusion that there is no infection, while a positive test leads to a confirmatory analysis using the Western blot (WB) or Immunoblot (IB). These tests require laboratory facilities.

1.2.4.2 Rapid diagnostic tests (RDTs) do not require any special equipment. RDTs for the detection of antibodies to HIV 1 and 2 provide access to serological status information for people who are unable to use conventional screening methods. Screening is carried out on serum, plasma or whole blood. For hard-to-reach populations, it can be carried out using dried blood on a sero-buvard. A negative result from an initial RDT excludes HIV infection, except in the case of recent exposure dating back less than 3 months (primary infection). A positive result must be confirmed by a second RDT. The sensitivity of RDTs currently on the market is considered equivalent to that of 3eme generation ELISA tests.

1.2.4.3 Techniques for identifying plasma viral RNA by PCR or RT-PCR allow early diagnosis. They can be used to diagnose primary infection (exposure to HIV less than 3 weeks ago) and mother-to-child transmission (children under 18 months old). Primary infection is a time of high contagiousness: sexual contamination and post-partum contamination in breast-feeding women, with the risk of MTCT (high viral load in breast milk), hence the importance of screening.

HIV molecular diagnostics can be used to confirm the status of blood donors in the seroconversion phase, with the PCR being positive from day 11.

1.2.4.4 HIV self-tests

Self-detection of HIV-antibodies from blood or gingival fluid is an additional tool in the fight against delayed diagnosis. The WHO has published new guidelines on HIV self-testing to improve access to diagnosis. Self-testing is a way of reaching more people with undiagnosed infection.

1.2.4.5 Infant screening and diagnosis

Serological screening is not sufficient for infants, and virological screening must be carried out at 6 weeks of age, or even at birth, to detect the presence of HIV in infants born to seropositive mothers. New techniques are now available that enable screening to be carried out at the point of care, with results returned within a day (EID technologies).

1.2.5 Treatment

Antiretrovirals make up the therapeutic arsenal against HIV, which is gradually being expanded. Around twenty antiretroviral drugs are available in 2006, and their aim is to interfere with different mechanisms: on the one hand, the HIV enzymes needed for replication and, on the other, the mechanisms by which HIV enters the cell.

Thanks to the tritherapies used since 1996, deaths from AIDS have fallen significantly wherever these new treatments have been available. In the United States, for example, the widespread use of tritherapies has reduced the number of deaths each year from 49,000 in 1995 to 15,807 in 2016.

These drugs are likely to have temporary or permanent side effects, which may lead to the discontinuation or, more importantly, modification of the treatment, bearing in mind that, when correctly taken, they are relatively effective.

The intestine plays an essential role in the immunopathogenesis of the human immunodeficiency virus (HIV). Decreased levels of zonulin are associated with increased mortality in HIV patients. Treatment with maraviroc (CCR5 receptor antagonist) and raltegravir (integrase inhibitor) increases zonulin[123] . These combined data suggest that the zonulin pathway in its innate immune function may protect against HIV infection.

Antiretrovirals: Research into HIV/AIDS is extremely important, and a great deal of research, studies and publications are produced on a regular basis. However, since the time between the conception of a molecule and its marketing authorisation varies between seven and twelve years in France ([124]), it is important to put into perspective the effects of announcements, some of which will not lead to any direct practical application in the fight against HIV/AIDS.

So the only drugs recognised as truly effective are the antiretrovirals that have been granted marketing authorisation.

Since December 2021 in France it has been possible, under certain conditions, to replace the daily treatment with an injection every two months combining two retrovirals (cabotegravir and rilpivirine).

Antiretrovirals are classified according to their area of action:

Reverse transcriptase inhibitors: Reverse transcriptase inhibitors prevent the synthesis of proviral DNA (i.e. the DNA that enables the virus to be duplicated) from viral RNA. This class includes :

Nucleoside inhibitors (NRTIs): NRTIs were the first class of antiretroviral drugs to be marketed in 1985. They include zidovudine (AZT) (synthesised in 1964), didanosine (ddI), zalcitabine (ddC), stavudine (d4T), lamivudine (3TC) (1989 and used from 1995), abacavir (ABC) and emtricitabine (FTC).

Mutations in the genome due to reverse transcriptase give HIV resistance to NRTIs, which can be crossed between several NRTIs. These compounds are all neutral or reducing, with the exception of AZT, which is an oxidant.

Non-nucleoside inhibitors (NNRTIs): NNRTIs are potent, highly selective inhibitors of HIV reverse transcriptase. They include nevirapine and efavirenz. They are only active against HIV-1. They are metabolised to phenols by oxidation.

Nucleotide analogues: Nucleotide analogues such as tenofovir, which came onto the market in 2002, are organophosphorus compounds.

Protease inhibitors: The protease inhibitor (PI) class of antiretroviral drugs was launched in 1996. They represented a major turning point in therapeutic strategies against the human immunodeficiency virus. They act by inhibiting the action of the viral protease that enables viral proteins to be cut and assembled, a process that is essential for producing infectious viruses. The result is virions that are unable to infect new cells. PIs are active against HIV-1 and HIV-2, and do not create cross-resistance with NRTIs or NNRTIs.

Integrase inhibitors: These inhibitors block the action of integrase, preventing the viral genome from binding to that of the target cell.

The molecules currently available are raltegravir, marketed under the brand name Isentress, elvitegravir and dolutegravir, marketed under the brand name Tivicay.

Fusion inhibitors: Fusion-lysis inhibitors intervene at the start of the HIV replication cycle, by blocking HIV surface proteins or disrupting the coreceptors of cells targeted by HIV.

Several products are being studied and, in 2009, only enfuvirtide and maraviroc received marketing authorisation.

Choice of treatment: Since the early 1990s, a number of different tritherapies have been developed, which can be prescribed according to clinical stage, CD4+ T-cell count and viral load. Antiretroviral treatment currently comprises three drugs, generally two nucleoside reverse transcriptase inhibitors, combined with a protease inhibitor or a non-nucleoside reverse transcriptase inhibitor, or sometimes a third nucleoside reverse transcriptase inhibitor (tritherapies). A fusion inhibitor may also be used.

When treated for the first time, almost all patients achieve an undetectable plasma viral load within the first six months. This initial treatment must be as simple and as well tolerated as possible. Non-adherence to treatment is the main cause of therapeutic failure.

Although antiretroviral treatments are highly effective when followed correctly, HIV is still present in the body. Only its multiplication is slowed down. While it was once thought that the blood and sperm of infected people remained contagious, the Partner2 study, presented at the 22[e] International AIDS Conference, has shown that a person with an undetectable viral load does not transmit the virus.

Although research is very active and a number of candidate vaccines exist, one of which has produced encouraging results in terms of the feasibility of developing a vaccine[130] , there is no truly effective vaccine against this virus. Condoms offer simple, effective protection during sexual intercourse; since the early 2010s, the prophylactic use of the antiretroviral agents Emtricitabine/tenofovir (Truvada) has also proved effective against transmission during sexual intercourse, particularly in populations at particular risk (homosexual men, sex workers). Blood donations are subject to donor selection, systematic screening and specific treatment. Prevention also involves the use of single-use syringes on all occasions, particularly in cases of intravenous drug use or substitution treatment.

Despite the widespread dissemination of information about the disease and prevention, some people nevertheless engage in risky behaviour (see article on AIDS risk-taking), which requires preventive action.

1.2.7 Prevention

There are several ways to prevent HIV/AIDS, to protect yourself and others. Depending on the time of year, sexual practices, their frequency,... 1 or other of these methods may be preferred or more appropriate. These include :

- Voluntary screening. The rapid test or TROD (test rapide d'orientation diagnostique), using just a drop of blood; the self-test, which can be carried out at home, also using a drop of blood taken from the fingertip; the Elisa test, known as the 4[e] generation test, which requires a blood sample to be taken.

- Wearing protection during sex. It is a physical barrier against HIV and other sexually transmitted infections (STIs). Condoms are the only way to protect against HIV and other STIs. There are two types of condom: male and female.

- Loyalty;

- PrEP is recommended for all seronegative adults at high risk of contracting

HIV. More specifically:

- Men who have sex with men or transgender people;
- Sex workers ;
- People who come from countries with large numbers of HIV-infected people;
- Partners of a seropositive person who has not reached an undetectable viral load and who have :

o Had sex without a condom

o Had several different partners

o Has several episodes of STI

o Used Post-Exposure Treatment or PET (emergency treatment)

o Uses substances during sex (chemsex)

- Treating seropositive people with antiretroviral drugs helps to prevent the transmission of HIV to a non-infected person. By taking their treatment correctly, a seropositive person can reduce their viral load to the point where it becomes very low. A seropositive person with an undetectable viral load thanks to their treatment does not transmit HIV sexually. It is therefore a means of preventing the spread of the virus.
- Awareness campaigns in high-risk areas.

2. Poliovirus/Poliomyelitis

2.1 Polio virus

2.1.1 **Definition: Polioviruses, the** causative agents of poliomyelitis, belong to the Enterovirus genus. Following numerous studies since the 1930s and particularly in the 1950s, the human poliovirus has become a model of choice for the study of the molecular biology of animal RNA viruses.

2.1.2 Structure.

They are single-stranded linear RNA viruses of positive polarity, i.e. their genome is in the form of an RNA molecule of the same polarity as the messenger RNA. **Structure of the poliovirus** capsid. The positions of the VP1, VP2 and VP3 proteins are shown for a pentamer consisting of 5 protomers. Each protomer is bounded by an axis of symmetry 5, an axis of symmetry 3 and two axes of symmetry 2.

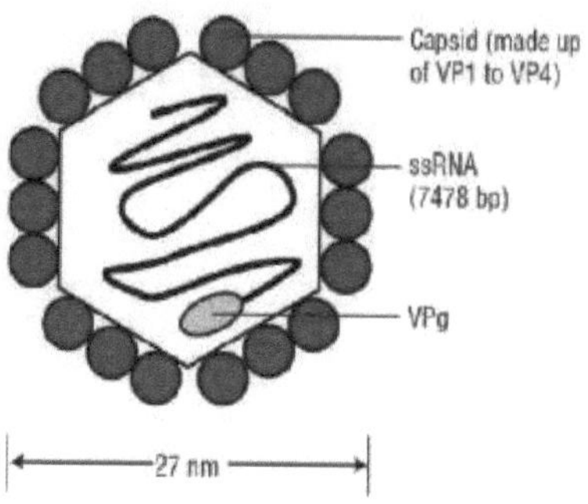

Figure 24: Structure of the poliomyelitis virus

2.1.3 Systematic position

Type	Virus
Group	Group IV
Family	Picornaviridae
Type	Enterovirus
Species	
Poliovirus	

2.1.4 Mode of infection

Polioviruses are transmitted orally and multiply in the tonsils and lymphoid tissue of the digestive tract. The non-enveloped virus is resistant to digestive solvents. The incubation period is 10 to 14 days.

Poliovirus is absorbed onto the surface of the host cell via the PVR receptor, a specific receptor for this virus. This receptor is present on the membrane of many cell types, but poliovirus can only multiply in cells of the anterior horn of the spinal cord. The virus penetrates the host cell by injecting its genome directly into it. In humans, the poliovirus limits its multiplication to cells in the pharynx, intestine and nerve cells, even though it reproduces in other tissues in the laboratory. However, it spares intestinal cells, which are ideal for the polio virus to multiply.

Expression and replication of the viral genome take place within the cytoplasm, resulting in the formation of numerous viral particles. During these processes, the cellular machinery is diverted to the benefit of the virus. The viral particles are released by cell lysis.

Polioviruses are relatively stable viruses: they remain inactive for a long time after pasteurisation. The only known natural reservoir for polioviruses is man.

2.1.5 Multiplication

The positive-polarity **genome** is translated directly into a large polyprotein, which is then cleaved to give rise to three proteins, P1, P2 and P3. The maturation of these viral proteins involves a number of cascade cleavages.

2.1.6 Distribution

Humans are the only known natural host of the virus. It can, however, infect other primates in certain circumstances. What is sometimes called *pig polio* is caused by a distinct virus, *encephalomyelitis enzootica suum*, the etiological agent of Teschen disease. Similarly, mouse polio is caused by *Theiler's murine encephalomyelitis* virus. Experiments with these animal viruses have been used to understand the action of poliovirus in humans.

In the XXe century, thanks to vaccinations, the virus declined and then disappeared in most countries. At the beginning of the 21st^e century, the *wild human poliovirus* was officially endemic only in Afghanistan, Nigeria and Pakistan, where it has never been defeated. At the end of 2016, it appeared to have disappeared from Nigeria and was on the verge of disappearing from Pakistan (one of its refuges from vaccination pressure, where it remains endemic). Epidemiologists believe that if Pakistan eradicates polio, Afghanistan should follow shortly afterwards. Yet polio cases are steadily falling in Pakistan (306 cases in 2014, 54 in 2015, 20 in 2016 and according to the figures available at the end of 2017: eight in 2017). What's more, blood tests show that human immunity to this virus has never been higher in Pakistan (even in children aged between 6 and 11 months), following years of vaccination campaigns. It was thought that in the absence of a large number of infected children, the virus was disappearing from Pakistan and should no longer be detectable in humans in 2018 or 2019. However, a search for evidence of the presence of the virus in the environment carried out in 2017 showed that the virus is still present in sewers and waste water pits almost everywhere in Pakistan, particularly in places where it was thought to have disappeared. As this is the first time this type of research has been carried out, we have no point of comparison. Are there healthy carriers, or are there patients who have escaped the statistics (for example, children from the Pashtun minority, who are marginalised in the country)? 16% of analyses from 53 sampling sites across Pakistan are positive, while very few patients are detected by doctors. Another hypothesis is that this phenomenon is normal before the disappearance of a virus of this type, which despite its presence could disappear fairly quickly if it no longer finds children to infect.

On 25 August 2020, the World Health Organisation (WHO) declared polio "eradicated" from the African continent, after four consecutive years with no reported cases and massive efforts to vaccinate children. Only two countries are still infected with the wild poliovirus: Afghanistan, with 29 cases in 2020, and Pakistan, with 58 cases.

On 11 June 2021, the WHO declared that the disease had disappeared from the Philippines following a vaccination campaign.

2.2 Poliomyelitis :

2.2.1 Definition: Polio **is a** contagious disease that can be prevented by vaccination. It **is** caused by poliovirus types 1, 2 or 3. It is transmitted from person to person and through contaminated food and water.

2.2.2 Geographical distribution

Worldwide, polio (wild virus) has been declared officially eradicated in the Americas (1994), the Western Pacific (2000), Europe (2002), South-East Asia including India (2014), Africa (2020), with only Pakistan and Afghanistan remaining. In France, the last case of indigenous poliomyelitis dates back to 1989, and the last imported case was declared in 1995[2] .

The wild type 2 virus was certified eradicated in 2015 (last case in 1999), and the wild type 3 in 2019 (last case in 2012). As of this date, virus type 1 is the only type of wild virus still in circulation.

However, with the decline in the number of cases caused by wild viruses, the number of cases caused by vaccine viruses is now in the majority.

In 2018, 33 cases worldwide were reported using the wild virus (21 in Afghanistan and 12 in Pakistan), and 104 using the vaccine virus (77 in Africa and 27 in Asia).

In 2019, there will be 176 cases of the wild virus, 29 in Afghanistan and 147 in Pakistan, and 378 cases of the vaccine virus in 19 countries.

In 2020, there were 140 cases of the wild virus, 56 in Afghanistan and 84 in Pakistan, and 1,039 cases of the vaccine virus in 26 countries, most of them in Africa.

2.2.3 Diagnosis

Additional examinations

Analysis of the cerebrospinal fluid (CSF) collected by lumbar puncture revealed a clear fluid, moderate hypercytosis (high number of cells in a body fluid), predominantly lymphocytes, normal glycorachy, normal or moderately increased proteinorachy, indicating aseptic meningitis. Repeated 15 days later, the examination generally shows a decrease in the number of cells and an increase in proteinorachy. However, the use of lumbar puncture is not without risk, particularly in epidemic periods.

Poliomyelitis serology is sensitive and early, but diagnosis of certainty requires direct evidence of the poliovirus on a throat swab, in the faeces or in the CSF. This is only carried out exceptionally in endemic areas, as it is costly and not essential. However, it is necessary in doubtful cases, particularly in areas where the disease has disappeared. Identification of the viral genetic material by polymerase chain reaction also makes it possible to distinguish wild strains from the vaccine strains used for oral vaccination. This distinction is important because for every reported case of paralytic poliomyelitis, there are an estimated 200 to 3,000 other asymptomatic but contagious cases.

Differential diagnosis

The differential diagnosis is very difficult in non-paralytic forms, as the disease is often mistaken for a common nasopharyngeal or digestive infection.

When a meningeal syndrome is present, it does not differ from other viral meningitis.

Paralytic poliomyelitis is clinically suspected in the acute onset of flaccid paralysis of one or more limbs with reduction or abolition of osteotendinous reflexes, without any sensory or cognitive impairment. Other clinical features of acute flaccid paralysis that may point to poliomyelitis are its asymmetric nature, rapid progression, association with fever and the occurrence of sequelae. The diagnosis is easily made in natives of endemic areas, more rarely in non-immune travellers.

The diagnosis of poliomyelitis requires the exclusion of another cause, notably inflammatory (Guillain-Barre syndrome, acute transverse myelitis) or mechanical (medullary or radicular compression, trauma related to an intramuscular injection). Other types of enterovirus may cause poliomyelitis-like paralysis. Arboviruses may also be involved. West Nile fever is one example[174,175]. There are also bacterial diseases such as diphtheria and botulism. Other diagnoses include Kugelberg-Welander syndrome and myotonic dystrophy.

2.2.4 Treatment

Medicine does not recognise a cure for poliomyelitis. Extra-neurological forms and aseptic meningitis, if diagnosed as such, can only be treated symptomatically. In the case of paralytic poliomyelitis, the aims of therapeutic management are to alleviate symptoms, speed up recovery and prevent complications. Treatment includes analgesics to combat pain, antibiotics to treat bacterial superinfections, moderate physical exercise and a suitable diet. Treatment often requires a prolonged convalescence, combined with physical rehabilitation, the use of orthoses, orthopaedic shoes, mobility aids (wheelchairs, canes, etc.) and, in some cases, orthopaedic surgery.

2.2.5 Prophylaxis

Most of the disease is transmitted by the faeco-oral route, so one of the best ways to prevent it is to improve hygiene.

The only possible preventive medical action is to vaccinate children.

There are two types of vaccine:

• The inactive injectable polio vaccine (IPV) was developed in the 1950s and contains the 3 strains of poliovirus involved in the disease. It induces a good level of immunity, and requires several injections with regular booster doses. However, its cost has long limited its spread in certain developed countries, including France;

• The oral poliomyelitis vaccine (OPV) was also developed in the 1950s. It also contains the 3 strains of poliovirus. To date, this vaccine has been the preferred tool for eradicating poliomyelitis, because it is easy to use, is not injected and rapidly confers good immunity. It is also much more affordable. On the other hand, it is more difficult to store and its genetic instability is a disadvantage (possible cause of the rare cases of paralytic polio associated with the vaccine).

All infants must be vaccinated with 3 injections (at 2 months, 4 months and 11 months). Boosters, which are essential for protection against the disease, are then given during childhood (at 11 and 13) and continue into adulthood (at 25 and 45). The vaccine is paid for.

3. SARS-COV2/COVID-19

3.1 SARS COV2

3.1.1 Definition

Sars-CoV-2 is the official name of the new coronavirus identified on 9 January 2020 in the city of Wuhan, capital of Hubei province in China. Initially called the Wuhan coronavirus or 2019-nCoV, the name was proposed by the *International Committee on Taxonomy of Viruses,* the body responsible for classifying viruses.

It is the etiological agent of the infectious pneumonia epidemic that spread in China and around the world from the end of December 2019. This disease was named Covid-19 by the WHO on 11 February 2020.

3.1.2 Structure :

Virus protein structure

Representation of the peplomere, glycoprotein of SARS-CoV-2 (PDB: 6VSB), forming the spicule. The whole protein is a homotrimer, of which only one monomer is detailed here, the rest of the trimer being shown in grey. Some parts of the actual structure are not shown. The protein domains are shown, from the N-terminus (letter N) to the C-terminus (letter C): N-terminal domain; ACE2 receptor-binding domain; general structure; central helix; the connector domain that anchors the peplomere to the lipid envelope of the virus. Disulphide bonds and carbohydrates in red. The "floor" of the virus' lipid membrane

Like other coronaviruses, SARS-CoV-2 has four structural proteins:

1. **S protein** (known as *spike* protein, *spicule* or *spicular protein*): forms the

peplomeres, protuberances of the "crown" characteristic of coronaviruses. These peplomeres play a key role in binding the virus to one or more receptors on the surface of a cell. Protein S appears to be one of the main determinants of SARS-CoV-2 tropism.

Publication of the viral genome has made it possible to model its three-dimensional structure. It has also been described at atomic level using cryogenic electron microscopy;

2. **Protein E** (envelope) ;
3. **Protein M** (membrane) ;
4. **Protein N** (nucleocapsid); this envelops and protects the viral RNA (the virus's genetic code).

Proteins S, E and M together make up the viral envelope.

Genome

The SARS-CoV-2 genome contains 11 genes recognising 15 *Open Reading Frames* (ORFs) enabling between 29 and 33 viral proteins to be produced after proteolysis. Initially, 29 viral proteins were identified. At the end of 2020, at least 4 additional viral proteins were proposed (ORF2b, ORF3c, ORF3d and ORF3d-2).

The SARS-CoV-2 genome has a cap at its 5' end and a polyadenylated tail at its 3' end:

• At its 5' end are the ORF1a and ORF1b genes encoding non-structural proteins (*NSPs*). ORF1a and ORF1b represent two-thirds of the genome and overlap slightly, so they are sometimes confused and referred to as ORF1ab. ORF1ab is considered to be a single gene encoding 16 proteins, several of which are enzymes that play an essential role in genome replication and expression;

• At its 3' end are the ten genes encoding structural and accessory proteins. SARS-CoV-2 has four genes specific to structural proteins: S, M, E and N. The N protein gene also integrates the ORFs for producing two accessory proteins: 9b and 9c. In addition, SARS-CoV-2 has six genes associated with accessory proteins: 3a, 6, 7a, 7b, 8 and 10. The ORF3a gene also encodes the bulk of the ORF3b protein and 3 other viral proteins (ORF3c, ORF3d and ORF3d-2). It has also been suggested that the ORF encoding protein S produces a second protein called ORF2b.

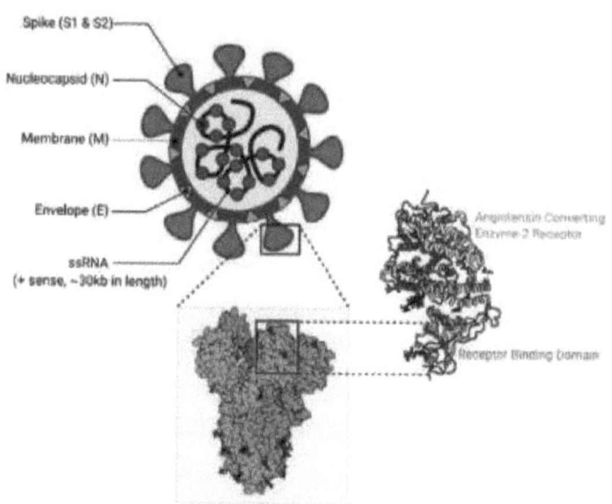

Figure 25: Structure of SARS Cov-2

The accessory proteins of SARS-CoV-2 diverge in part from those of SARS-CoV-1 :

• SARS-CoV-2 and SARS-CoV-1 share proteins 3a, 6, 7a, 7b, 9b and 9c;

• Proteins 3b and 8 of SARS-CoV-2 are different from proteins 3b, 8a and 8b of SARS-CoV-1. Along with protein N, ORF3b and ORF8 are thought to be the first targets of antibodies produced by B lymphocytes following infection with SARS-CoV-2. And this was long before antibodies were produced against fragments of protein S. This suggests that the ORF3b and ORF8 proteins are of major importance in pathogenesis;

• SARS-CoV-2 is thought to have 5 novel accessory proteins: 10, 2b, 3c, 3d and 3d-2.

3.1.3 Systematic position

Type	Virus
Kingdom	Riboviria
Regne	Orthornavirae
Branches	Pisuviricota
Class	Pisoniviricetes
Order	Nidovirales
Suborder	Cornidovirineae
Family	Coronaviridae
Subfamily	Orthocoronavirinae
Type	Betacoronavirus
Subgenre	Sarbecovirus
Species	SARSr-CoV

Form: SARS-CoV-2 ICTV

3.1.4 Types of infection

Possible modes of transmission for SARS-CoV-2 include contact, droplet and airborne, infected surfaces, feco-oral, blood, mother-to-child and animal-to-human. Infection with SARS-CoV-2 mainly causes moderate to severe respiratory disease, which can lead to death, while some people infected with the virus never develop symptoms.

J Transmission by contact and droplets

SARS-CoV-2 can be transmitted by direct, indirect or close contact with an infected person via infected secretions such as saliva and respiratory secretions or via respiratory droplets, which are expelled when an infected person coughs, sneezes, talks or sings (2-10). Respiratory droplets are >5-10 gm in diameter, whereas droplets <5 gm in diameter are called droplet nuclei or aerosols. Respiratory droplet transmission can occur when a person is in close contact (less than 1 metre) with an infected person who is showing respiratory symptoms (e.g. coughing or sneezing) or who is talking or singing; in these circumstances, it is possible for respiratory droplets containing the virus to reach the mouth, nose or eyes of a susceptible person and cause infection. Indirect transmission involving contact between a susceptible host and an infected object or surface may also be possible.

J Aerial transmission

Airborne transmission is defined as the spread of an infectious agent due to the dissemination of droplet nuclei (aerosols) that remain infectious when suspended in the air over long distances and for long periods of time. Airborne transmission of SARS-CoV-2 can occur during medical procedures that generate aerosols ("aerosol-generating acts"). WHO is actively discussing with the scientific community whether SARS-CoV-2 can also spread by air in the absence of aerosol-generating procedures, particularly in poorly ventilated farm settings.

3.1.5 Multiplication

After taking control of the infected cell, and while the Replicase-Transcriptase Complex (RTC) is replicating the viral genome, the ribosomes are mobilised to produce a series of structural viral proteins. These proteins assemble in the lumen (inside) of a compartment derived from the endoplasmic reticulum. This stage is called budding. First, a protein N (nucleocapsid) attaches itself to a copy of RNA and packs it into a protein M (membrane) which gives the virion its shape. S proteins are then incorporated. Protein M directs most of the protein-protein interactions required for virus assembly after binding to the nucleocapsid. Protein E contributes to the assembly and release of the virion from the infected cell, following the secretory pathway (Golgi apparatus, then secretory vesicles). The virion leaves the intracellular environment by exocytosis, and is ready to infect another cell.

3.1.6 Mutations and variants

Initially reputed to be stable, SARS-CoV-2 has turned out to be an extremely unstable virus. Mutations in the S protein are the ones that have been most mediated. The first major mutation is D614G, which favoured the infectivity of SARS-CoV-2. Since December 2020, the English variant has been distinguished by at least 17 modifications (mutations or deletions), all viral proteins combined, a record. The best-known mutation is N501Y, which improves the binding of RBD to the ACE2 receptor. The English variant doubles the infectivity of the virus. At the same time, South African and Brazilian variants have appeared, which share the N501Y mutation with the English variant. But these two variants above all contain mutations such as E484K, which weaken the effectiveness of antibodies to first-generation vaccines and facilitate reinfections with SARS-CoV-2.

In total, at the end of March 2021, of the 1,273 codons in protein S, 28 mutations (2%) have been identified that are spreading:

- 11 of these (40%) concern NTD: mutations at codons 18, 69, 70, 80, 138, 144, 215, 222, 241, 242 and 243;
- 6 of these (20%) concern the RBD: mutations at codons 417, 439, 452, 477, 484 and 501;
- 4 concern SD1 and SD2: mutations at codons 570, 614, 677 and 681.

NTD is the most unstable fragment of protein S, in other words the one that mutates most rapidly. The next major mutation in NTD is expected to occur at codon 248. This mutation could further weaken the antibodies of first-generation vaccines (AstraZeneca, Pfizer, Moderna...).

Apart from the S protein, at the end of March 2021, SARS-CoV-2 concentrates strong mutations in :

- ORF9c: of the 73 codons of this protein, seven (10%) are in the process of evolution: 194, 199, 202, 203, 204, 205 and 220. The ORF9c protein is essential for SARS-CoV-2 to deregulate the genes of infected cells;
- ORF9b: of its 97 codons, four (4%) mutate quite strongly: 10, 16, 32 and 70. In addition to its anti-interferon activity, ORF9b blocks apoptosis (cell self-destruction);
- ORF8: of its 121 codons, seven (6%) already have advanced mutations: 27, 52, 68, 73, 84 and 92. The ORF8 protein enables SARS-CoV-2 to block the presentation of an antigen by the MHC-I of infected cells, thereby delaying the

adaptive immune response;

• ORF3b: of the 156 codons in this protein, only four show strong evolution: 171, 172, 174 and 223. But ORF3b is the SARS-CoV-2 protein with the strongest anti-interferon activity. Above all, ORF3b is truncated. The ORF3b that encodes it contains four stop codons that should be skipped as mutations occur, gradually enabling this protein to exert increasingly strong anti-interferon activity.

Finally, there are thousands of variants of SARS-CoV-2, some of which have a negative impact on human health, diagnostics and vaccines. On the whole, the SARS-CoV-2 variants that stand out are those that are more contagious, more deadly and more resistant to first-generation vaccines and diagnostics. In March 2021, a variant identified in Brittany has the particularity of not being detectable by PCR testing.

3.2 COVID-19

3.2.1 Definition

Covid-19 refers to "*Coronavirus Disease 2019*", the disease caused by a virus in the *Coronaviridae* family, SARS-CoV-2. This infectious disease is a zoonosis, the origin of which is still debated, which emerged in December 2019 in the city of Wuhan, in China's Hubei province. It spread rapidly, first throughout China, then abroad, causing a global epidemic.

Covid-19 is a respiratory disease that can be fatal in patients weakened by age or another chronic illness. It is transmitted by close contact with infected people. The disease could also be transmitted by asymptomatic patients, but scientific data is lacking to confirm this with certainty.

3.2.2 Geographical distribution

Since the declaration of the COVID-19 outbreak on 31 December 2020, a total of 191,127 cases of infection, including 7,807 fatal cases (a fatality rate of 4.1%), have been reported worldwide, according to figures available as of 18 March 2020. As of 18 March 2020, 161 countries, territories or regions and one international transport vessel had reported laboratory-confirmed cases of COVID-19.

This statistic shows the number of people worldwide who will have died from the coronavirus (COVID-19) by 11 February 2022, depending on the country. Out of a total of more than 405 million infections linked to the virus worldwide, 5.8 million people have died to date, including more than 130,000 in France.

With 906,000 deaths, it is the United States that has claimed the most victims. Although little reliable scientific information is currently available, the fatality rate of the virus is estimated at between 2% and 3%. Furthermore, health professionals report that the majority of victims of the coronavirus (COVID-19) were elderly (people aged over 80 are most at risk) or suffering from previous pathologies. COVID-19 does not therefore systematically cause the death of those infected: in fact, most of them recover.

3.2.3 Diagnosis

Currently, screening for **coronavirus** infection is based on a PCR (polymerase chain reaction) test, which shows whether or not the virus has RNA (ribonucleic acid) in a nasopharyngeal swab inserted deep into the nasal cavity.

3.2.4 Treatment

There is currently no treatment capable of eradicating the virus. The only care given to patients is to treat the symptoms. Researchers around the world are exploring numerous avenues to find an antiviral drug or a vaccine, with no convincing results to date. Antibiotics are ineffective against viral infections, as are certain traditional plant- and food-based remedies.

In around 80% of cases, patients recover spontaneously, without needing any special treatment. The most serious cases are treated in intensive care units in hospital, where they are closely monitored.

3.2.5 Prophylaxis

To prevent the spread of COVID-19, follow these recommendations: Keep a safe distance from everyone (at least 1 metre), including people who do not appear to be ill.

- Wear a mask in public places, especially indoors or when physical distance is not possible.
- Open, well-ventilated areas are preferable to enclosed spaces. Open a window if you are indoors.
- Wash your hands frequently with soap and water or a hydro-alcoholic solution.
- Get vaccinated as soon as you can, following local vaccination recommendations.
- If you cough or sneeze, cover your nose and mouth with your elbow or a handkerchief.
- Stay at home if you don't feel well.

4. Hepatitis B virus/Hepatitis B

1.1. Hepatitis B virus

1.1.1. Definition

Hepatitis B is a liver disease caused by a DNA **virus** in the hepadnaviridae family. Left undiagnosed, it can progress to cirrhosis and even liver cancer. With 350 million people affected **worldwide**, it is one of the most common chronic diseases.

4.1.2 Structure

The **hepatitis B virus** (HBV) is a 42 nm enveloped **virus** belonging to the Hepadnaviridae family. Its genome is a double-stranded circular DNA of 3,200 nucleotides, comprising one long strand and one short strand. It is a small genome with a partially overlapping reading frame. It is the smallest of the human DNA viruses. It encodes just 4 genes:

1. **The C gene,** with a pre-C zone, for the capsid or core, consisting of the HBc antigen with a molecular weight of 21,000 (21 kDa).

2. **The S gene**, with a pre-S1 zone and a pre-S2 zone, for the envelope, made up of HBs antigen (s for surface). This HBs antigen comes in three forms: small, medium and large, of 24, 33 and 39 kDa, depending on whether it comes from the expression of the S gene, pre-S2 + S, or pre-S1 + pre-S2 + S.

3. **The P gene,** for polymerase, more precisely DNA polymerase, 90 kDa.

4. **The X gene**, with its poorly understood transactivating functions, may be involved in HBV-induced cancerogenesis.

Four genes in such a small genome implies a particular, very economical organisation: the viral DNA is circular, with 2 strands over 50 to 80% of its length, and above all, its 3 reading frames are used, so that the genes overlap, making the best use of the virus's limited genetic capacity. The P gene, the longest of the four, corresponding to three quarters of the genome, thus completely overlaps the S gene and partially the C and X genes.

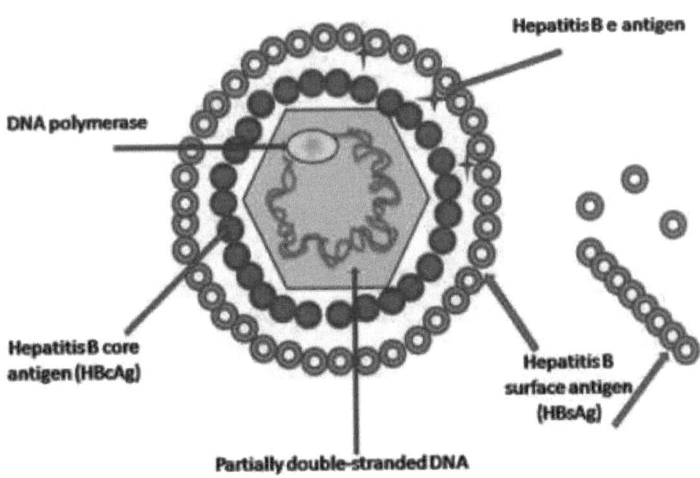

Figure 26: Structure of the hepatitis B virus

4.1.3 Systematic position of the hepatitis B virus

Type   Virus

Group	Group VII
Family	Hepadnaviridae
Type	Orthohepadnavirus
Species :	*hepatitis B virus*

4.1.4 Mode of infection

The virus is transmitted via biological fluids and secretions. The main modes of transmission are sexual intercourse, injections by drug users, unsafe blood transfusions, transmission from mother to child during childbirth and close contact with an infected person. Once in the bloodstream, the virus reaches the liver and multiplies in its cells, the hepatocytes. The immune system destroys the infected cells, causing inflammation of the liver.

4.1.5 Multiplication

Man is the only natural host. Infection of chimpanzees and other captive primates has been attributed mainly to human contact. However, specific HBV variants have also been observed circulating in these animals. These raise interesting questions about the origin of HBV, but this primate infection does not currently play a role in human epidemiology.

Permissive cells are hepatocytes, although viral DNA has been found in small quantities in extra-hepatic sites, monocytes, B lymphocytes, CD4+ and CD8+ T lymphocytes. This is probably related to the graft reinfections observed after liver transplantation, particularly in patients with severe chronic hepatitis.

However, *in vitro* multiplication of HBV, which can be obtained in primary cultures of hepatocytes, or in certain continuous lines of cells retaining the properties of liver cells, appears to be very limited compared with what is observed *in vivo* in humans.

4.1.6 Historical background

The first recorded epidemic of hepatitis B was observed by Lurman in 1885: an outbreak of smallpox was reported in Bremen in 1883 and 1,289 shipyard workers were inoculated with the lymph of others; after several weeks, and up to eight months later, 191 of the inoculated workers fell ill and developed an icon, and were diagnosed with serum hepatitis. The other employees, inoculated with different batches of lymph, remained in good health. Lurman's publication, now considered a classic example of epidemiological research, proved that lymphatic contamination was the cause of the epidemic. Later, many similar cases were reported following the introduction of hypodermic needles in 1909, which were used over and over again to administer Salvarsan for the treatment of syphilis.

The existence of a virus was suspected as early as 1947, but the virus was not discovered until 1963 when Baruch Blumberg, a geneticist working at the

American NIH at the time, detected an unusual reaction between the serum of polytransfused individuals and that of an Australian aborigine. He believed he had discovered a new lipoprotein in the aboriginal population, which he named the "Australia" antigen (later known as the hepatitis B surface antigen, or HBsAg). In 1967, Blumberg published an article[19] showing the relationship between this antigen and hepatitis. The name HBs subsequently came to be used to designate this antigen. For his discovery of the antigen and for designing the first generation of hepatitis vaccines, Blumberg was awarded the Nobel Prize for Medicine in 1976.

Virus particles were observed using an electron microscope in 1970. The virus genome was sequenced in 1979 and the first vaccines were tested in 1980. Different genotypes (at least nine, graded from A to I) were identified, and their geographical distribution studied. The study in 2018 of 12 genomes of the virus in bones dating back 800 to 4,500 years, combined with less complete results on 304 other ancient skeletons, traces the evolution of the virus and its relationship with hepatitis B viruses in the great apes. The geographical distribution of the ancient genomes is not identical to the current distribution, but it is compatible with what is known about human migrations during the Bronze and Iron Ages. Genetic drift is estimated at $8\text{-}15 \times 10^{-6}$ nucleotide substitutions per site per year, which implies an age of between 8,600 and 20,900 years for the root of the hepatitis B virus gene tree.

4.2 Hepatitis B

4.2.1 Definition

Hepatitis B is a viral hepatitis caused by infection with the hepatitis B virus (HBV), leading to inflammation of the liver.

The symptoms of ai guc disease are essentially inflammation of the liver, with or without icterus, and digestive disorders with nausea and vomiting. At this stage, revolution is often benign, although hepatitis B is the most serious form of viral hepatitis.

4.2.2 Geographical distribution

Hepatitis B is found throughout the world, but particularly in Asia, the Middle East, Africa and parts of America. In Switzerland, around 0.5% of the population is infected with the hepatitis B virus, compared with a global average of around 3.5%. Around forty cases of acute hepatitis B are reported each year in Switzerland, and the trend is downwards, with men by far the most affected (75% of cases). The majority of infections (around 55%) occur in the 35-60 age group.

4.2.3 Diagnosis

The seriousness of HBV infection is essentially linked to the possible progression from chronic hepatitis to cirrhosis and hepatocarcinoma. Diagnosis is largely based on serology.

J Clinical diagnosis

The clinical examination of a chronic hepatitis B carrier is normal, with the exception of moderate asthenia in some cases. In the case of chronic active hepatitis, certain symptoms may appear. These include a low-grade fever, enlargement of the liver and/or spleen (hepatomegaly and/or splenomegaly), icteric attacks (flu-like symptoms: headache, joint and muscle pain, but also nausea, diarrhoea and dark urine) and extrahepatic manifestations due to deposits of immune complexes (e.g. periarteritis nodosa).

In cases of cirrhosis, clinical signs of hepatocellular insufficiency and portal hypertension are observed.

J Immunological diagnosis

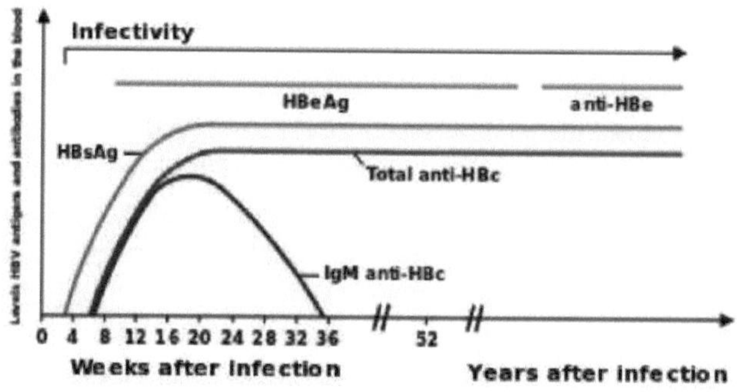

Figure 27: IgM evolution in HBsAg, HBeAg and HBcAg infection

Hepatitis B antigens and antibodies detectable in the blood during chronic infection (figure 27).

The specific diagnosis of HBV viral hepatitis is based on the detection of certain serum markers:

- antibodies: IgG anti-HBs, IgG anti-HBe, IgM and IgG anti-HBc ;
- antigenes: HBs and HBe ;
. HBV DNA.

Antigen detection is carried out using RIA (Radio Immuno Assay) tests. Serum HBV DNA is detected using molecular hybridisation techniques (PCR).

4.2.4 Treatment

There is **no specific treatment for acute hepatitis B'** (the period following infection). Patients should rest, eat low-fat foods and avoid alcohol or any medication that could be toxic to a liver already weakened by the hepatitis B virus. There are medicines indicated for the treatment of chronic hepatitis B. But

most people with chronic hepatitis B do not need these drug treatments. Regular medical check-ups are all that is needed.

When the hepatitis B virus is very active, or when the liver shows signs of inflammation or fibrosis, the doctor will consider treatment against the hepatitis B virus. In severe forms of chronic hepatitis B, and particularly in people where it is diagnosed very late, hospitalisation may be necessary to treat symptoms, introduce treatment and monitor progress.

4.2.5 Prophylaxis

Vaccination against hepatitis B is preferably recommended for infants, using a hexavalent combination vaccine, at 2, 4 and 12 months of age. Vaccination is also recommended between the ages of 11 and 15 for people not yet vaccinated against hepatitis B, as well as for at-risk groups at any age, including healthcare workers and drug users.

Not sharing needles or getting a tattoo in countries where the disease is widespread reduces the risk of contracting it.

People who change sexual partners frequently or who have several partners in the same period should therefore talk to their doctor or another health professional about HIV and other sexually transmitted infections, and ask if any tests are necessary.

5. Lassa/Lassa virus

5.1 Lassa virus

5.1.1 Definition: **Lassa fever is** a sudden-onset hemorrhagic fever caused by an arenavirus called the Lassa virus, closely related to the Ebola virus disease, first described in 1969 in the town of Lassa, in Borno State, Nigeria.

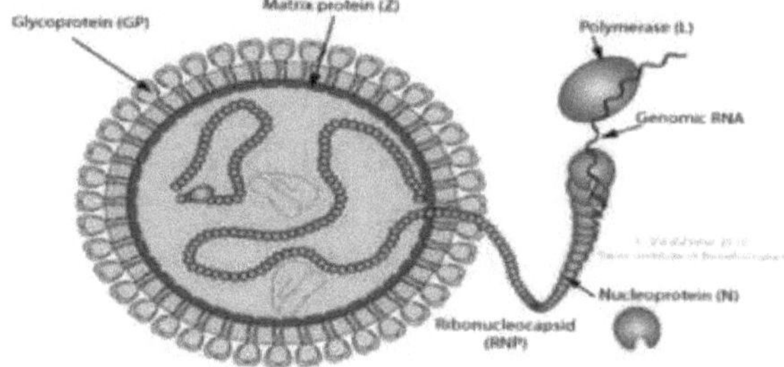

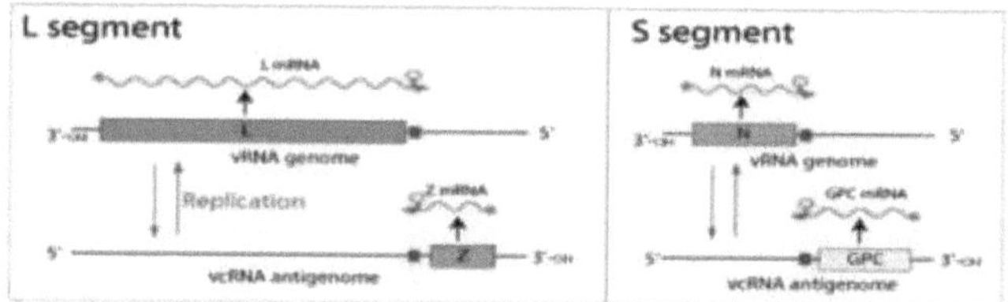

Figure 28: Structure of the Lassa virus

5.1.2 Structure :

The genome of this envelope virus consists of two segments of RNA, each encoding two proteins, one in each direction, for a total of four. The large segment, 7 kilobases long, encodes, in the positive direction, a small 11 kDa **Z** zinc finger protein that regulates transcription and replication; in the negative direction, it encodes the 200 kDa **L** RNA-dependent RNA polymerase. The small segment, 3.4 kb long, encodes, in the positive direction, the 75 kDa **GP** precursor of the surface glycoproteins, a precursor which is then cleaved by proteolysis into two envelope glycoproteins **GP1** and **GP2** which bind to the *alpha-* dystroglycan (en) receptor (a-DG) and allow the virus to penetrate the host cell; in the negative direction, it encodes the 63 kDa nucleoprotein **NP**.

5.1.3 Systematic position :

Type Virus

Domaine Riboviria

Negarnaviricota **branch**

Sub-embr. Polyploviricotina

Class Ellioviricetes

Order Bunyavirales

Family Arenaviridae

Type Mammarenavirus

Species
Mammarenavirus Lassa **ICTV**

5.1.4 Mode of infection

The Lassa virus enters the host cell via a-DG receptors. Receptor recognition depends on a particular modification of an a-DG ose by specific glycosyltransferases. Specific variants of the genes encoding these proteins are particularly present in West Africa, where Lassa fever is endemic. The presence of an aliphatic amino acid residue at position 260 on the GP1 glycoprotein is essential for the latter's affinity for l'a-DG. The nature of the preceding residue (at position 259) also appears to be a determining factor, since all viruses of the *Arenavirus* genus with a high affinity for a-DG have a residue with an aromatic side chain - tyrosine or phenylalanine - at this position.

Unlike most enveloped viruses, which use clathrin-lined cavities to penetrate their host and bind to their receptor in a pH-dependent manner, the Lassa virus follows an endocytosis pathway that is independent of clathrin, caveolin, dynamin and actin. Once inside the cell, viral particles are rapidly transferred to endosomes through the vesicular circulation. Fusion of the viral envelope with the vesicular membrane occurs via the interaction of the viral glycoprotein GP2 with the lysosomal protein LAMP1 (en) under the effect of the acidic pH of the endosome. Understanding the mechanisms underlying the conformational changes induced by the binding of the viral glycoprotein to its receptor and by membrane fusion is the focus of research aimed at developing a vaccine against Lassa fever.

Given the biological danger it represents, the Lassa virus can only be handled in a P4 or BSL-4 laboratory.

5.1.5 Multiplication

The replication of the Lassa virus occurs in successive stages which contribute to immunosuppression. Given the ambisense nature of the genome of this virus, its replication first produces a large number of copies of a complementary RNA viral genome used to massively express the proteins encoded in the negative direction, i.e. nucleoprotein NP and RNA-dependent RNA polymerase L. The viral genome is then transcribed identically to the original virus from this complementary RNA genome. The viral genome is then transcribed identically to the original virus from this complementary genome to complete viral replication, allowing massive expression of the proteins encoded in the positive direction, i.e. the zinc finger protein Z and the precursor of glycoprotein GP7, which still has to be cleaved into GP1 and GP2. The latter are thus produced last, which further delays identification of the virus by the host immune system, causing the immunosuppression seen in cases of Lassa fever.

5.2 Lassa fever

5.2.1 Definition : Lassa fever is a viral haemorrhagic fever ждиё lasting from one to four weeks that occurs in West Africa. The **Lassa** virus is transmitted to humans by contact with food or household items contaminated with rodent urine or droppings.

5.2.2 History: The first cases were reported in the 1950s, but it was not until 1969 that the virus was isolated from a nurse in the Nigerian town of Lassa.

The conflicts affecting some of the countries in the endemic zone are causing population movements that are conducive to epidemics. Sierra Leone, for example, experienced an epidemic of exceptional proportions between 1996 and 1997, due to a civil war.

5.2.3 Geographical distribution: Lassa fever occurs mainly in West Africa. It is endemic in Nigeria, Guinea (Conakry), Liberia and Sierra Leone. According to the WHO, it also affects other West African countries.

5.2.4 Epidemiology

In endemic areas, up to 50% of the population is thought to be infected with the disease. Epidemiological studies have identified between 300,000 and 500,000 cases a year in West African countries[5] . Of these 300,000 to 500,000 cases, 5,000 to 6,000 individuals die each year from Lassa fever[6] . The case fatality rate is around 1%, but reaches 15% in hospitalised patients. In pregnant women, the mortality rate is 30%, and the foetus dies in 85% of cases.

5.2.5 Diagnosis :

Because the symptoms of Lassa fever are highly variable and not very specific, clinical diagnosis is often difficult, especially in the early stages of the disease. It is difficult to distinguish Lassa fever from other viral haemorrhagic fevers, such as Ebola virus disease, and from many other fever-producing diseases, including malaria, shigellosis, typhoid fever and yellow fever.

Diagnosis of certainty requires tests to be carried out only in reference laboratories. Laboratory samples may present a biological risk and require extremely careful handling.

Infection with the Lassa virus can only be diagnosed with certainty by the following laboratory tests:

* enzyme-linked immunosorbent assay (ELISA);
* antigen detection;
* gene amplification preceded by reverse transcription (RT-PCR);
* virus isolation in cell culture.

5.2.6 Treatment

The only effective treatment at present is intravenous injection of ribavirin, an antiviral agent used in particular to treat hepatitis C[4] . To be effective, treatment must be administered at an early stage of the disease[4] . Prescribed for the first 6 days, it can reduce the rate of hepatitis C infection by 90%. This treatment can have a number of adverse effects, in particular severe anemia. Fortunately, these side effects are reversible once the treatment is stopped.

The low specificity of early symptoms makes diagnosis difficult in the early days of the disease. This is the main limitation of this treatment, since the antiviral is only effective in the early stages of the disease. The disease therefore remains difficult to treat.

J **Project in progress**

Many laboratories are currently looking for a vaccine against Lassa fever. In particular, research is being carried out by Institut Pasteur and INSERM researchers in the P4 laboratory at the Merieux-Pasteur Research Centre in Lyon, France.

Researchers have isolated proteins on the surface of the virus that activate antibody production. However, there are significant variations, up to 20%, in these proteins. As a result, immune responses can be very different. To produce an effective vaccine, we need to find the "common denominator" for all these glycoproteins, which stimulate the production of antibodies that act in all cases.

There is little hope, but these proteins are the only solutions currently known, opening the way to the development of a possible vaccine.

5.2.7 Prevention

Preventing Lassa fever involves promoting good "community hygiene" to prevent rodents from entering homes. Effective measures include storing cereals and foodstuffs in rodent-proof containers, disposing of rubbish away from homes, keeping them clean and keeping cats away from homes.

Mastomys rats are so abundant in endemic areas that it is impossible to completely eliminate them from the environment. Families must always take care to avoid any contact with the blood and biological fluids of a sick person.

In the medical environment, staff must always take the usual precautions to prevent and control healthcare-associated infections when caring for patients, whatever the presumed diagnosis. These precautions include basic hand hygiene, respiratory hygiene, the wearing of personal protective equipment (to protect against splashes or other contact with contaminated materials), safe injections and funeral rites.

Health workers caring for suspected or confirmed cases of Lassa fever should take additional infection control measures to avoid contact with the patient's blood or body fluids and contaminated surfaces or materials such as clothing and bed linen. When in close contact with patients (less than one metre away), they must wear face protection (face shield or surgical mask and goggles), a clean, non-sterile, long-sleeved gown and gloves (sterile for certain medical procedures).

Laboratory staff are also at risk. Samples taken from humans or animals to test for Lassa virus infection must be handled by qualified personnel and analysed in laboratories using the strictest possible containment conditions.

6. Ebola virus/Ebola virus disease

6.1 Ebola virus

6.1.1 Definition

The **Ebola virus is** the infectious agent that causes frequently haemorrhagic fevers in humans and other primates - Ebola virus disease - and has been the cause of epidemics of historic proportions and severity.

6.1.2 Structure :

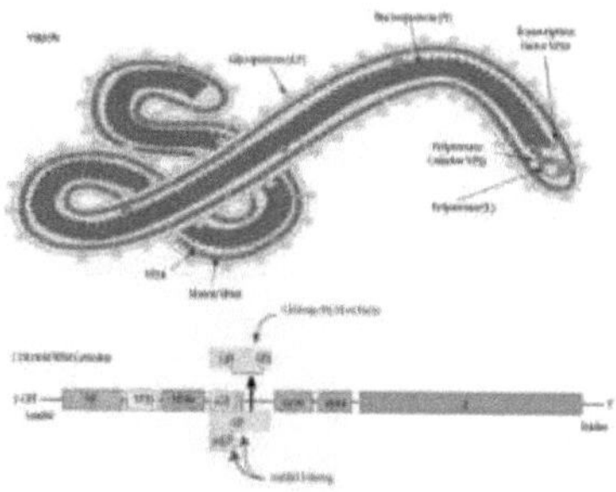

Figure 30: Structure of the Ebola virus

The Ebola virus can be linear or branched, ranging in length from 0.8 to 1 gm, but can reach 14 gm by concatemerisation (formation of a long particle by concatenation of shorter particles), with a constant diameter of 80 nm. It has a helical nuclear capsid 20-30 nm in diameter. This is made up of NP and VP30 nucleoproteins, which in turn are enveloped by a helical matrix 40-50 nm in diameter, made up of VP24 and VP40 proteins and comprising 5 nm transverse striae. The matrix is in turn enveloped by a lipid membrane containing GP glycoproteins.

Figure 31: Organisation of the Ebola virus genome.

It has a 19-kilobase genome with an organisation characteristic of filoviruses.

6.1.3 Systematic position

Type   Virus
Domain   Riboviria
Branch   Negarnaviricota

Sub-embr.	Haploviricotina
Class	Monjiviricetes
Order	Mononegavirales
Family	Filoviridae
Type	Ebolavirus

Species: Ebolavirus Zaire ICTV

6.1.4 Historical background :

The Ebola virus was named after a river that flows past the town of Yambuku in northern Zaire (now the Democratic Republic of Congo). It was in the hospital there that the first case of Ebola haemorrhagic fever was identified in September 1976 by the head doctor of the Bumba health zone, Congolese doctor Ngoy Mushola, who gave the first full clinical description, heralding a first epidemic that would affect 318 people and kill 280.

The Belgian doctor, Peter Piot from the Institute of Tropical Medicine in Antwerp, who has long been wrongly presented as the discoverer of the Ebola virus, was part of the first laboratory team to work on what was later identified as a new virus. In his own words, he should have been more reactive when someone declared him "the discoverer of the virus".

The blood samples used to identify the virus, once attributed to Congolese researcher Jean-Jacques Muyembe, were in fact taken by the team made up of Dr^r Firmin Krubwa, from the University of Kinshasa, and Drs^r Gilbert Raffier and Jean-Francois Ruppol for the blood of convalescents, and by Dr^r Jacques Courteille, from the Clinique Ngaliema in Kinshasa, for the blood of a nurse who died of the disease, as attested in 2016 in the prestigious *Journal of Infectious Disease*, by the main players still alive in this first epidemic.

6.1.5 Mode of infection

For Munster, a specialist in Ebola and other dangerous pathogens, understanding the epidemiology of this virus (and in particular how a zoonotic virus passes from one species to another) involves an ecoepidemiological approach, because "logging, hunting and human settlement in pristine environments all play a role, bringing people into contact with the microbes that lurk there. Once a new infectious agent reaches humans, the forces of globalisation, urbanisation and mobility can spread it more rapidly than ever before".

J Between humans

Direct contact with the body fluids (blood, vomit, diarrhoea, sweat, suppurations, saliva, sperm, etc.) of an infected person is the main route of human-to-human contamination.

According to the WHO's conclusions as of October 2016, the most infectious fluids are currently blood, faeces and vomit. Conversely, the virus is not spread by coughing or sneezing, with a "rare or non-existent" risk according to the WHO's current observations: given the epidemiological data, including the latest outbreak, the spread patterns do not correspond to the characteristics of airborne diseases (measles or chickenpox viruses, or tuberculosis bacilli, for example).

The risk of spread among hospital staff is very high, particularly if equipment is not sterilised. In endemic areas, failure to observe hygiene and safety rules has led to the deaths of several doctors and nurses during epidemics, and encourages nosocomial contamination. Close contact, i.e. direct contact with the body fluids of an infected person, whether living or dead, is a source of contagion; the funeral rituals of certain Central African peoples, which consist of washing the body and then rinsing the hands in a communal basin, have also often encouraged the spread of the virus among the family and friends of the deceased.

J Between humans and animals

One hypothesis is that where fruit bats are particularly abundant, they could be a source of infection for other species, however from 2006 to 2017 "no one has isolated the virus living in bats, and no one knows how Ebola could pass from a bat to other mammals, including humans, or why this fatal jump is so unpredictable in time and geography". An as yet unidentified intermediate host could be involved.

Transmission to humans appears to be linked to the handling of primates (dead or alive) infected with the virus: the case of monkeys, probably of the genus *Cercopithecus*, sold as bushmeat on the markets in the Democratic Republic of Congo.

In the laboratory, non-human primates have been infected following exposure to aerosolised particles of the virus from pigs, but airborne transmission between primates has not been demonstrated. Pigs excreted the virus in nasopharyngeal secretions and faeces following experimental inoculation.

6.1.6 Multiplication

Fusion of the virion envelope with the plasma membrane of the host cell releases the nuclear capsid into the cytoplasm of the target cell. The L-dependent RNA polymerase partially denatures the genomic RNA and transcribes it into positively polarised messenger RNA, which is then translated into proteins. The Ebola virus L RNA polymerase binds to a unique promoter located at the 5' end of the viral genome. Gene expression then proceeds sequentially, with an increasing probability of interruption as the polymerase progresses along the genomic RNA strand to be transcribed: the first gene from the promoter is thus expressed more than the last gene at the 3' end. The order of the genes on the viral genome thus provides a simple but effective means of regulating their transcription: nucleoprotein NP, encoded by the first gene, is produced in greater quantity than polymerase L, encoded by the last gene.

The concentration of this nucleoproteine in the host cytosol determines when polymerase L switches from transcription - production of messenger RNA from genomic RNA - to viral replication - production of positive-polarity RNA antigenomes by integral replication of an original genomic RNA. These antigenomes are in turn transcribed into negative-polarity RNA viral genomes which interact with structural proteins previously translated from the viral RNA. Viral particles self-assemble from the newly produced proteins and genetic material near the cell membrane. They sprout outside the cell, covering themselves with a viral envelope derived from the plasma membrane, where the GP glycoproteins are inserted, releasing new virions ready to infect other cells.

6.2 Ebola virus disease

6.2.1 Definition

Ebola virus disease (formerly known as **Ebola** haemorrhagic fever) is a serious illness, often fatal in humans. The **virus** is transmitted to humans from wild animals and then spreads through human-to-human transmission. The average case fatality rate is around 50%.

6.2.2 Geographical distribution

Between 2014 and 2016, the resurgence of the **Ebola** virus disease **affected** almost 30,000 people in ten **countries**, mainly West Africans. Over 99% of fatal cases occurred in Guinea, Liberia and Sierra Leone.

6.2.3 Diagnosis

To confirm the **diagnosis**, blood samples must be sent to the country's National Reference Centre for Viral Haemorrhagic Fevers or to another biosafety level 4 (P4) laboratory, which is the only one authorised to carry out biological tests to determine whether or not the **Ebola** virus genome is present in the patient's blood.

Analysis of samples by PCR or ELISA must be carried out in a class III biological safety cabinet (glove box), with valid certification in a separate area of the laboratory.

• After inactivation, the samples can be removed from the glovebox and all other procedures can be carried out under biosafety level 2 conditions.

• Wear suitable personal protective equipment (PPE) to handle samples prior to inactivation: gloves, fitted masks such as N95 respirators with filtering facepieces (FFP) 3, powered air-purifying respirators (PAPR) if the fit test fails, full face shields and disposable waterproof gowns.

6.2.4 Treatment

The disease caused by the virus is fatal in 20% to 90% of cases. This large difference is due to the fact that the Ebola virus is particularly dangerous in Africa, where care is limited and difficult to provide for the population. Although there is no specific treatment for the virus, a number of symptomatic treatments (resuscitation, rehydration, transfusion, etc.) can prevent the patient from dying.

The first use of convalescent blood or serum as a treatment option, to take advantage of their antibodies and induce passive immunisation in transfused patients, was successfully attempted during the first Ebola epidemic in 1976 in Yambuku. A plasmapheresis programme was even implemented on that occasion, and was one of the recommendations of the International Commission deployed at the time by the Zairois government.

An experimental attenuated live vaccine has produced encouraging results in monkeys. It was administered in March 2009 to a researcher working on the virus, following a possible accidental contamination. The results were favourable. Since 2019, vaccination campaigns have been conducted in affected areas using this Ebola Virus Vaccine.

6.2.4 Epidemiology 2013-2015

This epidemic, which began in December 2013, is sometimes described as "atypical", because it is not under control[94] . In July 2014, it was developing in a worrying way in Guinea, Liberia and Sierra Leone. On 20 August 2014, 844 deaths were officially confirmed as due to the virus. An outbreak occurred in the district of Boende (an isolated region of Equateur province, in the Democratic Republic of Congo) and then died out. Another outbreak (with the first cases recorded in March 2014, unrelated to the other epidemic) spread across West Africa, becoming, according to the WHO, in just a few months "the largest and most complex outbreak since the virus was discovered in 1976. It is producing more cases and deaths than all previous outbreaks combined. Another feature of this outbreak is that it is spreading from one country to another, from Guinea to Sierra Leone and Liberia (by crossing land borders) and Nigeria (via a single air traveller)".

The epidemic began in March 2014 in Guinea (where it caused more than 2,500 deaths) and quickly spread to Liberia (4,800 deaths) and Sierra Leone (3,950 deaths). A number of neighbouring countries were affected but quickly brought the situation under control, including Mali (6 deaths) and Nigeria (8 deaths).

6.2.5 Prophylaxis

• The slaughter of infected animals using gloves and masks, with rigorous supervision of the burial or incineration of the carcasses, may be necessary to reduce the risk of transmission from animals to humans. Restricting or banning the movement of animals from infected farms to other areas may reduce the spread of the disease.

• The products (blood and meat) must be carefully cooked before being eaten.

• Communities affected by the Ebola virus must inform the population of the nature of the disease and the measures taken to contain the outbreak, including during funeral rites. People who have died from the infection should be buried quickly and safely.

• The imposition of quarantine, a ban on going to hospitals, the suspension of the practice of caring for the sick and of funerals, and the seclusion of the sick in separate huts that are disinfected (bleach at two-week intervals is sufficient) and sometimes burnt down after the death of their occupants, all help to contain epidemics. In the field, there is still no safer measure than wearing an air filter.

• Laboratory research must be carried out in biosafety level 4 containment facilities. Level 4 laboratories are fully self-contained and have a specialised ventilation system, an airlock for entry and exit, class III biological protection cabinets, etc. Sterilisation and decontamination procedures are rigorously applied and staff are trained in the use of these facilities. Sterilisation and decontamination procedures are rigorously applied and staff wear pressure suits.

7. Marburg/Marburg virus

7.1 Marburg virus

7.1.1 Definition

Marburg virus (MARV) is the causative agent of Marburg virus disease (MVD) in humans, with a rate of letalitis varying from 23 to 90%.

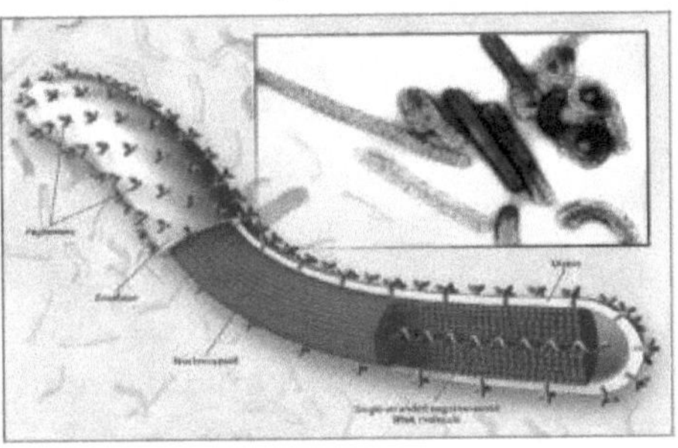

Figure 31: Structure of the Marburg virus

7.1.2 Structure.

Electron microscopy shows the **Marburg virus to** have a filamentous **structure** with a characteristic hooked end. The **virus** is endemic in equatorial Africa, where it is responsible for epidemic outbreaks that have so far been localised, but with fatalities ranging from 25 to 80 deaths.

7.1.3 Systematic position

Kingdom	Riboviria
Regne	Orthornavirae
Branch	Negarnaviricota
Sub-embr.	Haploviricotina
Class	Monjiviricetes
Order	Mononegavirales
Family	Filoviridae
Type	*Marburgvirus* ICTV, 2002

7.1.4 Mode of infection

Initially, human infection with MVD resulted from prolonged exposure to *Rousettus* bats inhabiting caves or mines.

MVD is spread by human-to-human transmission through direct contact (through the skin or lesioned mucous membranes) with the blood, secretions, organs or other bodily fluids of infected people, and with surfaces and materials (e.g. bedding, clothing) contaminated with these fluids.

Healthcare workers have frequently been infected when treating patients with suspected or confirmed MVD. This has occurred through close contact with patients when infection control precautions are not strictly applied. Transmission via contaminated injection equipment or needle sticks is associated with more severe disease, rapid deterioration and, possibly, a higher mortality rate.

Funeral rites involving direct contact with the body of the deceased can also contribute to the transmission of Marburg.

People who have contracted Marburg disease remain infectious for as long as their blood contains the virus.

7.1.5 Multiplication

Marburg is an enveloped virus with a negative-polarity RNA genome that causes fulminant hemorrhagic fever in humans and non-human primates. A single viral protein, glycoprotein (GP), is responsible for tropism and fusion of the viral membrane with the cell membrane, a mechanism that is more complex than it appears. During infection, GP interacts with several cellular proteins to promote adhesion, internalisation and transport of the viral particle into vesicles containing cellular proteins required for GP activation. GP catalyses fusion of the viral and cell membranes, enabling the viral genome to be delivered into the cell cytoplasm and initiating replication of the virus. In recent years, a number of discoveries have helped to improve our understanding of each of these different stages by identifying the cellular factors involved in infection.

However, the RNA of the Marburg virus is a molecule comparable to DNA, a kind of template that uses the information in DNA and transmits it to the cell's machinery, which in turn manufactures the proteins corresponding to the instructions. Marburg and Ebola are RNA viruses, which use this synthesis mechanism to replicate themselves.

7.2 Marburg virus disease

7.2.1 Definition

Marburg Virus Disease (MVM), formerly known as Marburg Virus Haemorrhagic Fever, is a serious and often fatal disease in humans. The virus causes severe viral haemorrhagic fever in humans. The average fatality rate for this disease is around 50%.

7.2.2 Geographical distribution

WHO epidemiological services report[16] in 6 African countries:

- Angola (2005, 329 deaths)
- Democratic Republic of Congo (1998 to 2000, 128 deaths)
- Kenya (1980, one death; 1987, one death)
- Uganda (summer 2007, two deaths; autumn 2014, one death)[17]
- Ethiopia: cases reported in 1984 and 1985, but censored by Mengistu's communist government, which was already facing a huge famine. Since then, the number of victims remains unknown.
- Gurnee: on 9 August 2021, the first case was detected in West Africa.

7.2.3 : Epidemiology :

In 1998, a larger epidemic affected 149 people near Durba, a town in the north-east of the Democratic Republic of Congo (DRC). More than 80% of these people died. Before 2000, cases were rare, almost all of them occurring in countries in eastern and southern Africa (South Africa, Kenya, Zimbabwe). In 2005, outbreaks in northern Angola affected more than 252 people, 227 of whom died (fatality rate: around 90%, comparable to the most severe Ebola epidemics).

Since October 2004, more than 400 cases have been observed in Angola by the World Health Organisation. While the first victims were children, adults were affected in spring 2005. On 6 October 2014, the Ugandan government announced a case of Marburg virus disease. On 17 October 2017, Uganda reported to the World Health Organisation the start of an epidemic of Marburg virus disease in the east of the country. Three people are reported to have died. Overall mortality is estimated at between 23% and 90%.

7.2.4 Diagnosis

It may be difficult, on the basis of clinical symptoms, to distinguish Marburg virus disease from other conditions such as malaria, typhoid fever, shigellosis, cholera and other viral haemorrhagic fevers. Symptoms caused by Marburg virus infection are confirmed using the following diagnostic methods:

- enzyme-linked immunosorbent assay (ELISA)
- antigen immunocapture test
- lockout test
- PCR with reverse transcriptase (RT-PCR)
- electron microscopy
- virus isolation in cell culture.

Samples taken from patients present an extreme biological hazard. Laboratory tests on samples that have not been inactivated must be carried out under maximum biological containment conditions. All biological samples must be protected in triple packaging when transported within the country and abroad.

7.2.5 Treatment

There is currently no vaccine or approved antiretroviral treatment for Marburg virus disease. However, supportive care - oral or intravenous rehydration - and the treatment of certain specific symptoms improve patient survival.

Monoclonal antibodies are being developed, and antiretrovirals such as Remdesivir and Favipiravir, which have been used in clinical trials for Ebola virus disease, may also be tested for Marburg virus disease, or be the subject of compassionate use or extended access.

In May 2020, the European Medicines Agency granted marketing authorisation for the Zabdeno (Ad26.ZEBOV) and Mvabea (MVA- BN-Filo) vaccines against Ebola virus disease. The Mvabea vaccine contains a virus known as *Vaccinia Ankara Bavarian Nordic*, which has been modified to produce four proteins from the Ebolavirus Zaire species and three other viruses in the same group (*filoviridae*). It is possible that this vaccine could provide protection against Marburg virus disease, but its theoretical efficacy has not been demonstrated in clinical trials.

7.2.4 Prophylaxis

Marburg fever is a very serious disease, but it can be avoided. It is essential to wear gloves and other appropriate protective clothing (mask in particular) when working in mines or cellars inhabited by colonies of flying foxes.

8. Influenza viruses

8.1 Flu Virus

8.1.1 Definition

The influenza virus (also known as influenzavirus or *Myxovirus influenzae*) is an RNA virus of the *Orthomyxoviridae* family. Its eight RNA molecules are housed in a protein capsid inside an envelope. The influenza virus is transmitted by air.

Influenza viruses are a group of four species of single-stranded RNA virus of negative polarity, each distinguished by a particular antigenic type: influenza A, influenza B, influenza C and influenza D[1] . Of these four antigenic types, type A is the most dangerous, type B presents fewer risks but is still likely to cause epidemics, and type C is generally associated with only minor symptoms. Type D is less widespread than the others and is not known to cause infections in humans. These four viruses have a segmented genome, eight segments for the first two and seven segments for the last two. The amino acid composition of influenza C and D viruses is similar at 50%, a rate similar to that observed between influenza A and B viruses; on the other hand, the rate of divergence between A and B viruses on the one hand and C and D viruses on the other is much higher.

8.1.2 Structure.

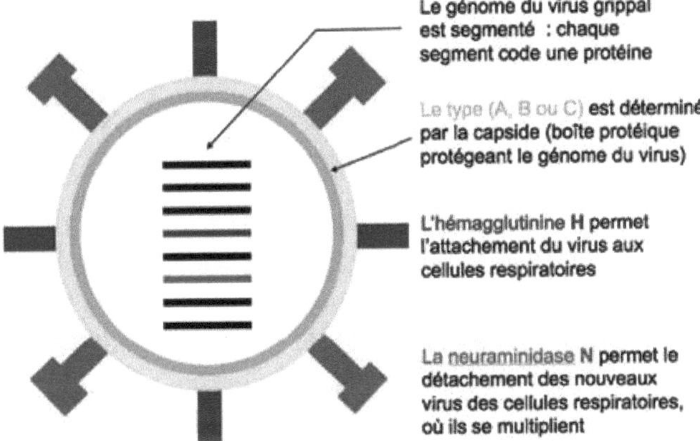

Figure 32: Structure of the influenza virus

The viral particle contains the 8 RNAs, encapsidated in viral proteins. The whole is surrounded by a layer of matrix proteins and enveloped in a cell-type membrane, forming a viral particle 80 to 120 nanometres in diameter.

The viral particle consists of a lipid envelope with spicules formed by surface glycoproteins. The A and B viruses have two surface glycoproteins: Hemagglutinin (H) and Neuraminidase (N).

Hemagglutinin, which represents around 40% of surface glycoproteins, is formed by the association of two subunits, HA1 and HA2, linked by a disulphide bridge. The combination of three HA monomers forms a hemagglutinin spicule on the surface of the viral particle. Hemagglutinin enables the virus to bind to the terminal sialic acid of the cells of the ciliated epithelium of the respiratory tract: it is highly immunogenic, inducing the production of antibodies, some of which may be neutralising. Hemagglutinin also promotes the fusion of viral and cellular membranes during the virus penetration phase.

8.1.3 Systematic position

Type Virus

Domaine   Riboviria

Negarnaviricota **branch**

Sub-embr.   Polyploviricotina

Class Insthoviricetes

Order   Articulavirales

Family   Orthomyxoviridae

Genus : Alphainfluenzavirus

Species: **influenza A virus**

Genus : Betainfluenzavirus

Species: **influenza B virus**

Genus: Gammainfluenzavirus

Species: **influenza C virus**

Genre : Deltainfluenzavirus

Species: **influenza D virus**

8.1.4 Mode of infection

Infected mammals generally transmit influenza through the air, by coughing or sneezing and expelling aerosols containing the virus, while infected birds transmit it through their droppings. Influenza can also be transmitted via saliva, nasal secretions, faeces and blood. Infections occur through contact with these bodily fluids or contaminated surfaces. Outside their host, influenza viruses can remain infectious for around a week at human body temperature (37°C), for more than 30 days at 0°C and indefinitely at very low temperatures - such as the lakes of north-eastern Siberia. They can be easily inactivated by disinfectants and detergents.

8.1.5 Multiplication

Influenza viruses, like all *Orthomyxoviridae*, replicate in the cell nucleus, a characteristic they share only with retroviruses. This is because they lack the enzymatic machinery needed to produce their own messenger RNA. They use cellular RNAs as primers to begin synthesising viral messenger RNA by a mechanism known as cap capture. Once in the nucleus, the RNA polymerase protein PB2 binds to the cap of the 5' end of a cellular pre-messenger RNA. The PA protein cleaves this RNA near its 5' end and uses this capped fragment as a primer to transcribe the rest of the viral RNA genome into viral messenger RNA. This method is necessary for the viral messenger RNA to have a cap so that it can be recognised by a ribosome and translated into a protein.

As the viral RNA-dependent RNA polymerase has no error correction function, it makes a nucleotide insertion error approximately every 10,000 insertions, which is about the length of the flu virus genome. Each new virion therefore statistically contains a mutation in its genome.

8.2 Flu

8.2.1 Definition

Influenza is a highly contagious acute respiratory infectious disease caused by viruses of the *Influenza* genus.

It evolves in epidemics, which can be worldwide when the virus responsible infects people who have never encountered it before. This is known as a **pandemic**.

Seasonal influenza occurs between November and April in the northern hemisphere and from April to September in the southern hemisphere. In tropical countries, the *influenza* virus circulates all year round.

8.2.2 Epidemics in the 19th century

The symptoms of human flu were clearly described by Hippocrates almost 2,400 years ago. Livy described brutal epidemics in ancient Rome, which in retrospect seem to be attributable to influenza. Since then, the virus has been responsible for numerous pandemics. Historical data on influenza is difficult to interpret, because the flu syndrome is also found in other epidemic diseases (diphtheria, bubonic plague, typhoid fever, dengue fever, typhus, hepatitis A). The first convincing observation dates back to 1580, with a pandemic that started in Asia and spread to Europe and Africa. More than eight thousand deaths were recorded in Rome and several Spanish cities were hit. Pandemics continued to occur sporadically throughout the seventeenth and eighteenth centuries[e] and [e68], and there was a widespread pandemic between 1830 and 1833 (a quarter of people exposed to the disease are thought to have been infected). It was not until the 1850s that a systematic description of epidemics was undertaken by the British scientist Theophilus Thompson.

8.2.3 Diagnostics

Chromatographic tests are also available for the qualitative detection of influenza viral antigens in samples prepared from respiratory specimens. These tests can give very rapid results (around 15 minutes) but tend to give false-negative results and cannot therefore replace more advanced tests. They are, however, widely used as presumptive tests to guide treatment quickly. Negative results are always confirmed by more sensitive and specific methods, such as PCR.

More and more laboratories are also using molecular biology techniques: extraction of viral RNA from the sample, followed by end-point RT-PCR or quantitative RT-PCR. These techniques enable a fairly rapid (less than two hours for extraction followed by quantitative RT-PCR) and reliable diagnosis, which also has the advantage of enabling an initial typing. RT-PCR can then be completed by sequencing the viral genome, essentially for epidemiological purposes.

8.2.4 Treatment and prophylaxis

Vaccines and medicines are available for the prophylaxis and treatment of influenza virus infections. Vaccines are made up of inactive or attenuated virions of the H1N1 or H3N2 subtypes of the human influenza A virus, as well as variants of the influenza B virus. As the antigenicity of wild viruses is constantly evolving, vaccines are reformulated each year using new strains. However, when the antigenicity of wild viruses does not match that of the strains used to produce these vaccines, the latter do not protect against these viruses, and when the antigenicities do match, mutants may also emerge that escape the antigenic coverage of the vaccines.

Among the drugs available for the treatment of human influenza, amantadine and rimantadine inhibit the release of virions from infected cells by interfering with the M2 matrix protein, while oseltamivir (marketed under the brand name *Tamiflu*), zanamivir and peramivir inhibit the release of virions by interfering with neuraminidase; peramivir tends to generate fewer mutants than zanamivir.

9. Measles virus

9.1 Measles virus

9.1.1 Definition

The **measles** *virus* (also known as MV) is a virus belonging to the *morbillivirus* genus of the *Paramyxovirus* family. It is a highly contagious virus that can cause serious, even fatal, problems such as encephalitis. Not everything is known about it, especially the cell receptors it targets. It is important to continue research in order to find better vaccines against this virus.

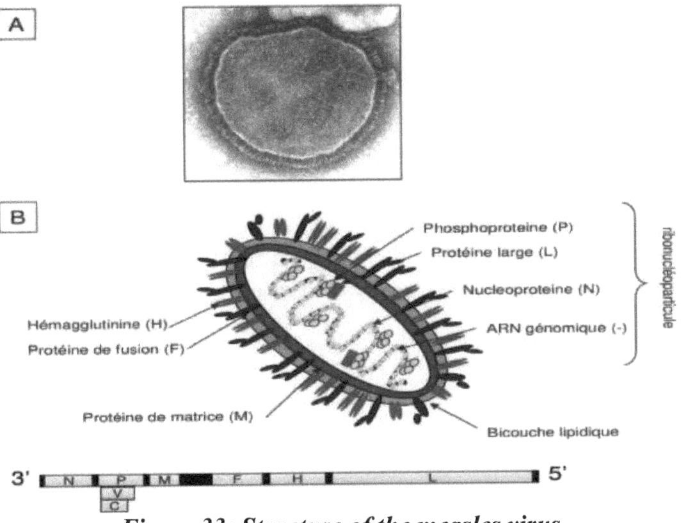

Figure 33: Structure of the measles virus

9.1.2 Structure.

MV is a virus with an envelope consisting of a double layer of lipids surrounding a helicoidally symmetrical nucleocapsid. Its diameter varies between 120 and 300 nm. Its genome is a non-segmented strand of negative RNA made up of 15,894 nucleotides forming 6 genes encoding at least 8 different proteins. A single promoter is present in the genome and is located at the 3' end. It is 56 nucleotides long. At the other end is the terminator, 40 nucleotides long. Each gene is separated from the next by 3 nucleotides that are not transcribed during transcription (Figure 33).

The virus has six genes[3] : F, H, L, M, N and P:

- M: matrix
- N: nucleocapsid
- P: phospholipid
- L: wide
- H: haemagglutinin
- F: fusion

9.1.3 Systematic position

Type   Virus
Domain   Riboviria
                Negarnaviricota

Branch
ous-embranch.   Haploviricotina
Class   Monjiviricetes

87

Order	Mononegavirales
Family	Paramyxoviridae
Subfamily	Orthoparamyxovirinae
Type	Morbillivirus

Measles morbillivirus species **ICTV**

9.1.4 Mode of infection

The measles virus is transmitted directly by airborne droplets of saliva. It can also be spread by direct contact with secretions from the nose or throat of infected people. The virus released in this way remains dangerous for at least thirty minutes and up to almost two hours, in a closed air environment (such as a doctor's surgery), or on objects and surfaces.

The virus begins to spread two to six days before the rash appears. The virus settles in the body during the incubation period. The virus is present in respiratory secretions from the end of the incubation period until the fifth day after the rash. The risk of transmission diminishes from the second day after the eruption.

It has been known since the 19th century[e] that this disease is "highly contagious". The contagious period begins five days before and lasts up to five days after the outbreak. The reproduction rate of measles (calculation of the average number of individuals that an infected person can infect for as long as he or she is contagious) in a non-immunised population is estimated at between 12 and 18, making it one of the most contagious diseases.

8.1.5 Multiplication

MV enters the body via the respiratory system. It first replicates in the immune cells (alveolar macrophages and dendritic cells) residing in the respiratory mucosa. These infected cells then travel to the lymph nodes, where they transmit the virus to the lymphocytes and monocytes present. The infection then spreads to the secondary lymphoid organs. In the late stages of infection, large numbers of infected immune cells circulate in the body and transmit the virus to the epithelial cells of the respiratory tract.

9.2 Measles

9.2.1 Definition

Measles (also sometimes called **first disease**) is an acute eruptive viral infection that mainly affects children aged 5-6 months and young adults. It is not benign and can lead to complications at any age. It is highly contagious, through the air: when coughing, close contact with infected people or through objects contaminated with secretions from the nose or throat (pacifiers, toys or handkerchiefs).

9.2.2 Diagnosis

Positive diagnosis: This is essentially clinical. It is based on the notion of contagion (cases already known in the entourage), the absence or inadequacy of vaccination, the presence of catarrh and Koplik's sign, and the characteristics of the exanthem (starting at the level of the head, and descending in a single shoot). In doubtful cases, serology can confirm the diagnosis through the presence of specific IgM antibodies that appear during the eruption. A negative serology within the first three days of eruption does not rule out the diagnosis. These specific IgM antibodies can be found in saliva between the first and sixth week following the onset of the disease.

The presence of specific immunoglobulin G helps in the diagnosis when 2 samples taken 10 days apart (between the acute phase and convalescence) show a significant rise (fourfold increase), otherwise its mere presence only indicates that the person has already been in contact with the disease-causing virus (either through previous infection or vaccination).

PCR techniques can confirm the diagnosis by sampling oculo-respiratory secretions, blood and saliva up to 5^e days after the start of the eruption. The virus can be isolated and genetically characterised, enabling the strain responsible to be identified for epidemiological purposes.

Differential diagnosis: Firstly, morbilliform exanthems: measles must be distinguished from measles, epidemic megalerythema, infectious mononucleosis, enterovirus, adenovirus infections, etc. Then there are other erythemas: infantile roseola, scarlet fever, Kawasaki, or an eruption of medicinal origin.

9.2.3 Treatment

It is a highly contagious disease and the doctor must immediately notify the health authorities so that measures can be taken. Measles usually lasts about ten days and then clears up on its own. The person is then immune to the measles virus for life. As it is a viral disease, antibiotics are not indicated and the fever, which can be very high, must be monitored.

9.2.4 Prophylaxis

The best form of prevention is still vaccination. This is what the health authorities regularly remind us when there is an upsurge in the disease. Two injections are needed to protect the child: the first at 12 months and the second between 16 and 18 months. If the child's vaccination schedule is not up to date, a catch-up vaccination is possible.

The vaccines used against measles are live attenuated virus vaccines whose virulence is reduced by the appearance of mutations that inactivate the virulence genes.

Before 1963, the world was plagued by waves of measles epidemics killing an average of 2.6 million people a year. The discovery of a vaccine in 1963 means that this disease can now be prevented, with the hope of one day eradicating it (the WHO believes this is possible). According to a WHO estimate, the widespread distribution of the vaccine prevented 20.4 million deaths between 2000 and 2016.

In terms of prevention, measles vaccination is compulsory for all children: a first dose of measles vaccine at the age of 12 months (or 9 months if living in a community). The compulsory vaccine, known as "MMR", also protects against two other diseases, measles and mumps (this is known as a trivalent vaccine). A second dose of this vaccine must be given in France between the ages of 13 and 24 months (or between 12 and 15 months if you live in a group). This second dose is not a booster, as the immunity obtained after the first dose is long-lasting. It is a catch-up for children who have not seroconverted against measles, mumps or rubella after the first injection. A single-dose vaccination produces immunity in 90-95% of people, while two doses produce immunity in over 98% of those vaccinated.

The duration of immunity is well in excess of several decades, and probably longer, as the calculation is based on measuring the persistence of specific immunoglobulins.

The vaccine is generally well tolerated, with less than 5% fever and a few cases of skin rash. There may be rare cases of a transient drop in the number of blood platelets.

9.2.5 Impact of vaccination (1980-2000)

Perceived as a benign disease in healthy people in developed countries, measles is in fact a very serious disease in children who are undernourished or living in poor hygiene conditions: in 1980, 2,600,000 people worldwide died from this disease.

Developing countries: In 1988, the WHO estimated that the vaccination programme had prevented 700,000 deaths from measles in developing countries in 1987. In 1997, in Africa, the number of cases and deaths fell by 40% (1990-1997).

However, there are still many cases of measles, both among children under nine months of age and because of sub-optimal vaccination coverage, with immunity not acquired after a single dose in 5% of cases. Africa remains the region with the highest incidence of measles (47.5 cases per 100,000 inhabitants per year) and the lowest vaccination coverage (57% of children under five not vaccinated against measles).

Developed countries: Finland was the first country to eliminate measles in

1993 (confirmed in 1996), with two-dose vaccination since 1982 and 96% vaccination coverage since 1991. A comparison of the results with countries that have only introduced a single-dose vaccination (either at the age of 2, or around the age of 5-6, or around the age of 11, depending on the country) has led to the introduction of a second dose in these countries.

United States: In the United States and in several Latin American countries, vaccination is compulsory before children can go to school. In some states, exemptions are possible for religious or philosophical reasons. Vaccination against measles became widespread from 1978. The overall number of cases fell by 90%, but epidemics persisted due to unvaccinated children of preschool age and vaccination failures (5% of single-dose vaccines).

In 1989-1991, a resurgence of measles occurred: the United States reported 55,000 cases and 123 deaths. They then adopted a two-dose strategy, also targeting preschool children (one dose at the age of 2, a second around the age of 56). In 1994, 958 cases of measles were reported, and indigenous transmission of the virus was interrupted in the years that followed.

France: In France, measles vaccination was included in the vaccination schedule in 1983, and has been compulsory since 2018. A first dose of vaccine is recommended at the age of one, and vaccination must be completed before the age of eighteen months. Vaccination coverage has been steadily increasing, with measles cases falling from almost 400,000 in 1987 to 44,000 in 1993. However, vaccination coverage stagnated at 80% in the 1990s, with disparities between departments, and the virus continues to circulate.

As vaccination coverage has increased, there has been a shift towards an earlier age, with measles occurring relatively more often in adolescents and young adults. These changes have led to modifications to the vaccination strategy, including the switch to two doses.

Modelling work has shown that a sub-optimal level of measles vaccination coverage is conducive to future outbreaks, confirming the need for a two-dose vaccination strategy, with the aim of achieving vaccination coverage of at least 95%. A second dose of the triple vaccine (measles-mumps-rubella) was therefore introduced in France in 1996, and reimbursed at 100% by the French social security system in 1999.

In 2000, compared with the end of the 1980s, the number of measles cases and deaths in France was reduced by more than 90%: from around 200,000 to 10,000 cases, and from 30 deaths to fewer than 3.

CHAPTER III: LABORATORY DIAGNOSTIC METHODS FOR VIRAL INFECTIONS

1. Introduction :

There are four stages in carrying out a diagnosis. The problem: choosing a central question, clarifying the concepts, proposing hypotheses for analysing the situation, proposing hypotheses for action to solve a problem or develop potential.

Viral diseases are diagnosed by clinical examination, biological tests, medical imaging, biopsies, lumbar puncture, in the case of infections that occur during epidemics, the presence of similar cases and, for certain infections, blood tests and cultures.

The choice between direct diagnosis (detection of a viral component) and indirect diagnosis (detection of antibodies directed against the virus) depends on the virus being tested for and the medical question being asked.

In animals and humans, antibodies indicate recent or past contact with a given antigen. Antibodies recognise a given antigen and are therefore specific to a given virus. Antibodies are generally tested in serum, hence the name serological test.

When the virus is cultivable, direct detection can be used to speed up diagnosis and achieve faster detection. The techniques used for the direct detection of antigens are the same as those used for the detection of antibodies, which have already been described.

Blood culture is the most common test used to diagnose an infection. It involves drawing blood from the patient and sending the blood sample to the laboratory. Additional tests may be carried out depending on the patient's signs and symptoms.

The ELISA (enzyme-linked immunosorbent assay) is a method used to quantitatively detect an anti-gene in a sample. There are four main types of ELISA: direct, indirect, competitive and sandwich. Each type is described below with a diagram illustrating the fagon in which the analytes and antibodies are bound and used.

The principle of rapid antibody detection by strip immunochromatography is identical. However, whereas tests to detect antigens do not require any prior processing of the sample, tests to detect antibodies require prior centrifugation of the blood tubes.

Real-time PCR is an accurate, rapid and cost-effective method. This method allows specific and sensitive detection of DNA and RNA and differentiation of numerous bacteria, viruses and parasites in various matrices (blood, stools, samples, etc).

2. Virus detection targets

Not all viral infections require laboratory intervention.

If justified by the clinical context or the severity of the symptoms, virological diagnosis uses either indirect or direct methods. Among the latter, molecular diagnosis is playing an increasingly important role.

Virus structure detected	Method	Current applications
Indirect diagnosis		
Anti-viral antibodies	Viral serologies	All viruses
Direct Diagnostics		
Complete viral particles	Viral culture	HHV, VZV, Enterovirus
Viral proteins	Rapid Direct Diagnosis	Rotavirus, Adenovirus, RSV, Influenza, Parainfluenza virus, HHV, VZV, CMV
Viral nucleic acids	Molecular biology (PCR)	neurotropic viruses (HHV-1...) HIV, HCV, HBV

3. Indirect diagnosis: viral serologies

To diagnose an ongoing infection, it is important to analyse **two consecutive samples taken** 10-15 days apart, in order to observe a significant change in antibody levels. The presence of IgM antibodies most often indicates a recent infection; IgG antibodies persist for a very long time.

When a person is serology positive for HIV and HCV for the first time, it is legally obligatory to verify this result on a second sample (serology and western blot).

4. 1. Sampling :

- Serum mainly, transported to the laboratory at **room temperature**.
- Other biological fluids: cerebrospinal fluid (CSF), amniotic fluid, pleural fluid, bronchoalveolar lavage fluid (BALF), etc. In these cases, a "home-made" adaptation of the serological technique is often necessary, which is unsatisfactory from a quality assurance point of view.

Blood samples taken in dry tubes and gelose tubes for serology: the blood is centrifuged and aliquots of serum are used to carry out the prescribed serologies.

Figure 34: Blood samples

3.2. Techniques:

- ELISA (Enzyme Linked Immunosorbent Assay) methods most commonly used
- Other, more "traditional" serological reactions, such as the complement fixation reaction
- Rapid tests (variable sensitivity, be careful!)
- HIV confirmation tests (western blot) and HVC (immunoblot): the different proteins of the virus are present, separately, on the membrane used as a reaction support. These tests make it possible to determine which viral proteins the antibodies are directed against.

3.3. Principle of the ELISA (enzyme linked immunosorbent assay) reaction :

- Formation of an antigen-antibody complex
- Serum is assumed to contain the antibodies sought
- Antigen (commercial reagent) adsorbed on a **plastic support** (96-well plate)

- Detection of the antigen-antibody complex by fixation of a branded **anti-human immunoglobulin antibody** (commercial reagent)
- by an enzyme (enzyme immuno-method)
- by a fluorescent molecule (immuno-fluorescence)

Figure 35: Enzyme-linked immunosorbent assay (ELISA) principle

94

Apparatus for serum distribution and management of ELISA methods: the serums are recognised by reading a barcode, then their distribution and the ELISA reaction are carried out automatically, in connection with the laboratory's computer system. The results are "validated" and interpreted individually by the technician and then the biologist, according to the values of the various controls and clinical information.

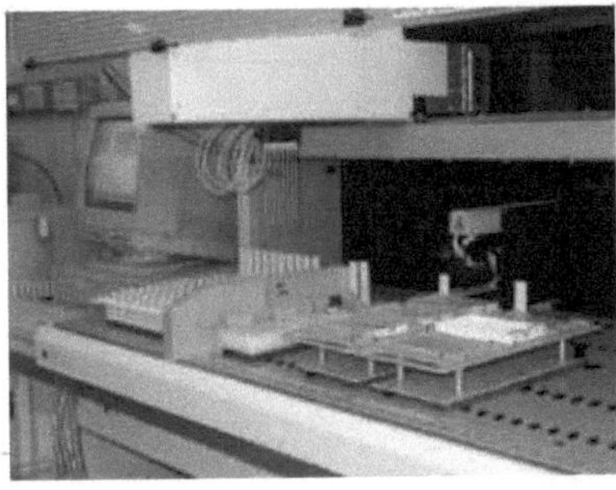

Figures 36 and 37: Equipment for serum distribution and ELISA method management

Aspect of an ELISA reaction plate at the end of handling: strongly coloured wells correspond to positive sera for the ELISA reaction. Optical densities are read automatically on a spectrophotometer.

Aspect of a complement fixation reaction plate at the end of the procedure: in the positive wells, the red blood cells are agglutinated together and form a precipitate at the bottom of the well.

Rapid HIV-1 and -2 serology test: this type of test is particularly useful in the event of a blood exposure accident. If the source patient's serum is positive, anti-retroviral treatment is started.

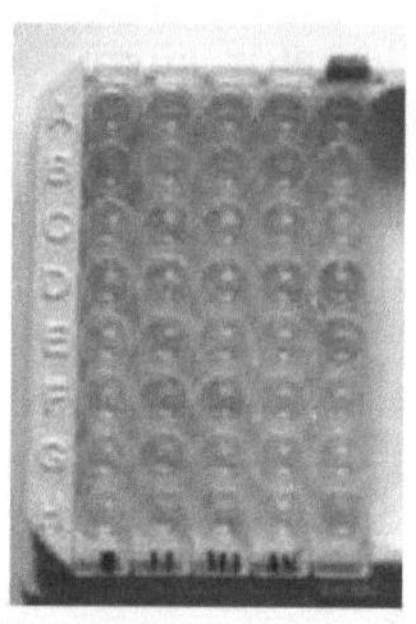

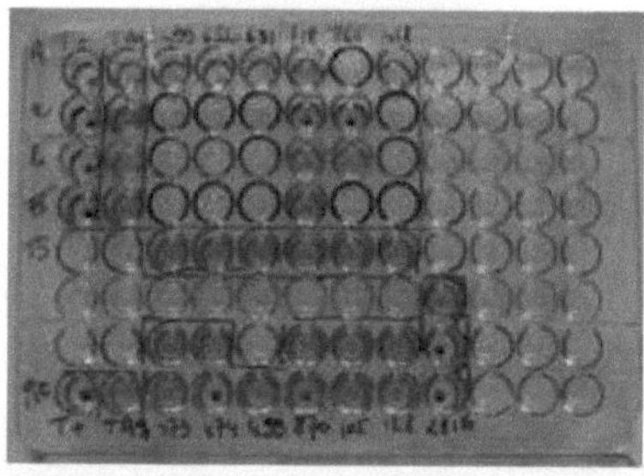

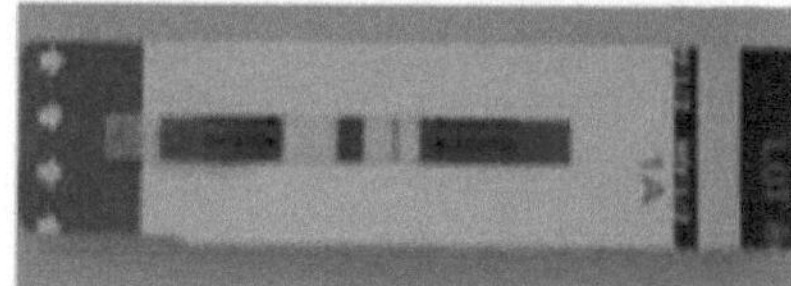

Figures 38-40: ELISA plates and HIV serology

very rapidly in exposed patients

Western blot strips at the end of the reaction: a coloured band appears when the serum tested contains antibodies recognising the antigen adsorbed at this point on the strip.

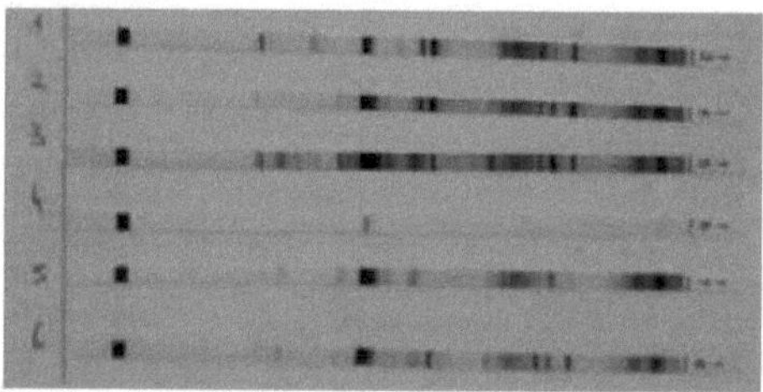

Figure 41: Western blot *strip* at the end of the reaction

- Indication: Evidence of more or less recent contact with a virus.
- Advantages :
- Automation
- Deadline for obtaining results: same day (4 hours) or next day
- Good sensitivity, excellent specificity (ELISA)

96

- Limits :
• Lower sensitivity in certain patients: infants and immunocompromised people.
• Sometimes difficult to interpret. Example: for all herpesviridae (HHV, CMV, EBV ...) IgM may or may not be present during reactivation.
- Indicative costs :
• HHV IgG and IgM serology: B70 + B70
• HVC serology: B70
• HIV serology (2 ELISA methods): B70
• HIV western blot: B180

4. Direct diagnosis

4.1.Collection :

The quality of the sample determines the result. There are three situations:
• Samples that do not require a transport medium: bring the sample to the laboratory
• in a sterile dry tube: **stool, urine, CSF, various liquid samples**
• in a tube containing 1 EDTA: whole blood for laboratory separation of blood mononuclear cells (testing for CMV, EBV, HIV, etc.)
• Samples taken with a sterile swab immersed in a transport medium supplied by the laboratory: **throat, vesicles, conjunctiva**

In both these cases, **send to the laboratory within 4 hours** (to preserve antigenicity, infectivity or the quality of the nucleic acids), - **at 4°C**, wrapping the sample tube in a bag of slides (to avoid bacterial proliferation) - special cases :
• **whole blood, at room temperature**
• **for molecular biology tests: EDTA+ + + tube** (no heparin, PCR inhibitor)
• Slide smear (the swab is expressed on the glass slide by the sampler): **transport dry, at room temperature**

Examples of samples: cerebrospinal fluid, bronchoalveolar lavage fluid, faeces, a sterile swab and a liquid transport medium.

4.2.Rapid" direct diagnostic methods

These methods deliver results in just a few hours.

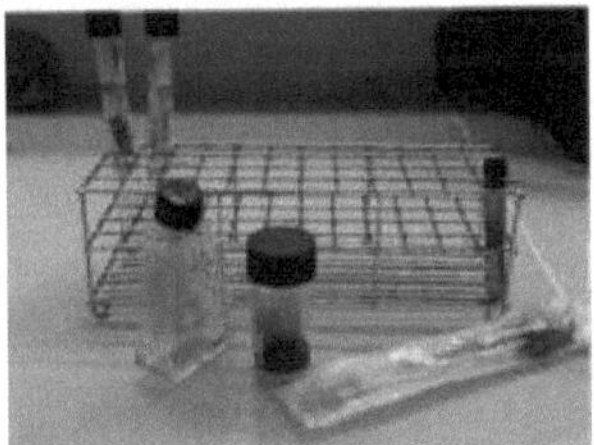

Figure 42: CSF samples

- Techniques:

• **Specific immunostaining for a virus, on a glass slide** in 1 or more "spots" (with an antibody directed against a viral protein), using a swab or blood mononuclear cells isolated in the laboratory. **Specify the virus being tested for**: only one immunostain is possible per "spot".

• Other possibilities: **agglutination of sensitised latex particles, ELISA-type method** for the automated detection of antigens, rapid enzyme-linked immunosorbent assay kits...

• Indications:

To enable specific and rapid treatment of patients in certain circumstances:

• pre-partum (HHV),

• Respiratory infections (influenza, respiratory syncytial virus, viruses, etc.) Parainfluenza, Adenovirus),

• Infant diarrhoea (Rotavirus, Adenovirus),

• search for sub-clinical CMV infection in an immunocompromised patient (pp65 antigenemia).

Agglutination test for Rotavirus in stools: stools, suspended in a buffer, are exposed to latex particles "sensitised" to the viral antigen. The result is negative in the case illustrated. If this were not the case, strict hygiene measures could limit the spread of the gastro-enteritis epidemic among hospitalized infants.

Specific immunofluorescence for HHV on a vaginal smear: infected cells show yellow-green fluorescence. The patient will be offered a cesarean delivery.

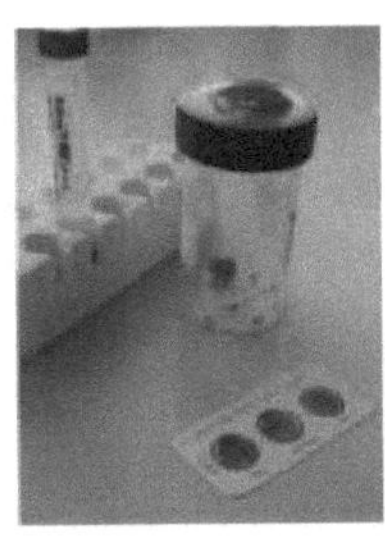

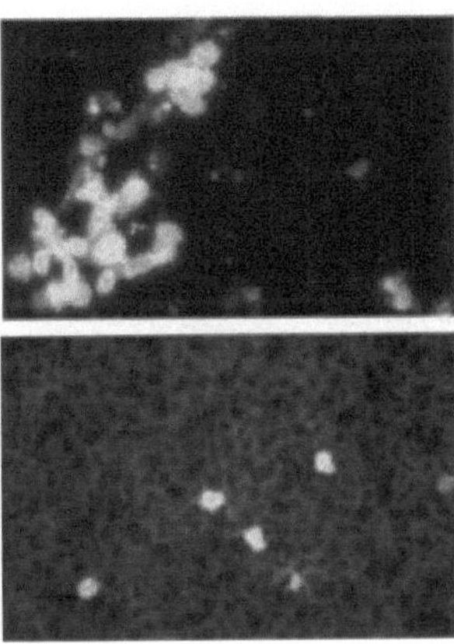

Figures 43-45: Agglutination test and HHV-specific immunofluorescence and CMV

CMV pp65 antigen positive: *the leukocytes marked by the antibody (yellow-green fluorescence) contain the 65kD phosphoproteine of CMV. The renal transplant patient will benefit from treatment against CMV, even if he or she does not have the antibody.*

does not yet present any clinical signs (pre-emptive treatment).

- Advantage :

speed of results: 20 minutes to 4 hours

- Limits :
- Lack of sensitivity + + +. A negative result does not exclude the diagnosis.
- Immunofluorescence: reading under the microscope depends on the observer's experience.

4.3. Direct culture diagnostic methods

Viruses are obligatory parasites of cells: they are cultivated in eukaryotic cells maintained in the laboratory, in the form of so-called continuous or semi-continuous lines.

- The different stages of diagnosis by viral culture

1. **Inoculation of** a cell line with the pathological sample.

2. Daily direct examination: **daily search for a cytopathic effect** (CPE) using an inverse light microscope.

3. If CPE: **staining of cells**, peeled and spread on a slide, for viral inclusions and cytoplasmic or nuclear alterations, using a light microscope.

4. **Specific immunostaining of** the virus on the slide, with reading by light or fluorescence microscopy.

Workstation for viral cultures: *laminar flow hood, disposable and sterile equipment.*

Observation of cell cultures *(tubes, plastic multi-well dishes and flasks) using an inverse optical microscope*

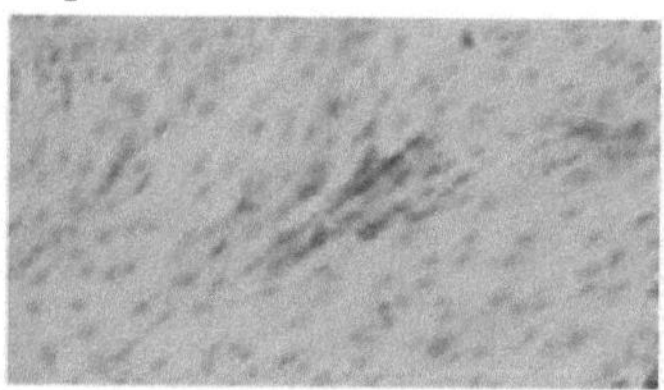

Cytopathic effect suggestive of HHV infection *in cultured MRC5 cells (human embryonic fibroblasts)*

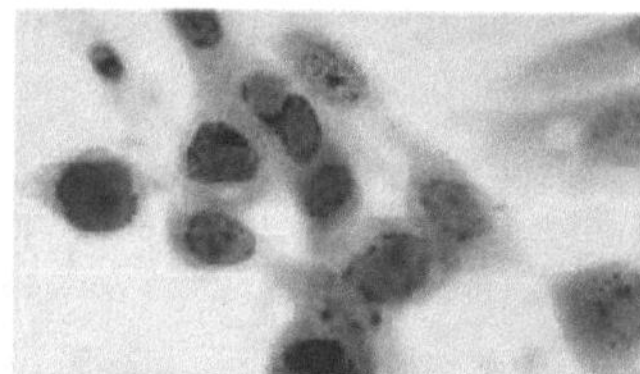

Cell staining showing the characteristics of this CPE: *disappearance of nucleoli, chromatin burst at the edges of the nuclear membrane, strongly*

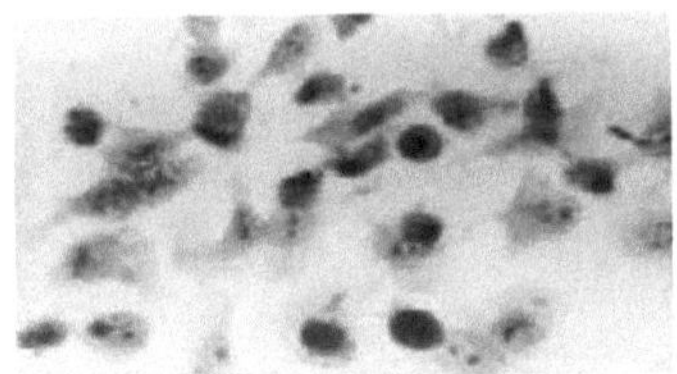

By comparison, ECP characteristic of an enterovirus culture: *intra-cytoplasmic inclusions and a hyperdense nucleus growing at the edge of the cell.*
Figures 46-50: Workstations, cytopathic effect and staining

- Indications:
- Isolating a virus from a biological sample.
- Proving the infectious nature of the virus.
- To characterise the phenotype of a viral strain (resistance to anti-virals, for example).

- Advantages :
- Good sensitivity + + +. Culture remains the benchmark against which all new methods are compared.
- It is essential for the availability of strains used, for example, in the development of vaccines (e.g. the influenza vaccine every year).
- Reasonable cost.

- Limits :
- Usual turnaround time for results: 48 hours to 10 days.
- Subjectivity of reading: the observer's experience +++.
- Culture is practised in few city laboratories.
- Some viruses cannot be cultivated outside highly specialised research laboratories.

- Cost :
- HHV culture on genital swab: B150
- enterovirus culture in stools: B150

4.4. Direct molecular diagnosis

In simple terms, three main methods dominate the routine: molecular hybridization and its variants (hybridization with signal amplification or bDNA, hybridization in search of mutations in the viral genome), gene amplification or PCR, and nucleotide sequencing.

Quantitative applications are becoming commonplace

- Techniques:

Hybridization: Direct detection of DNA or RNA on various supports (tubes,

plates, membranes, etc.).
Principle of molecular hybridization :

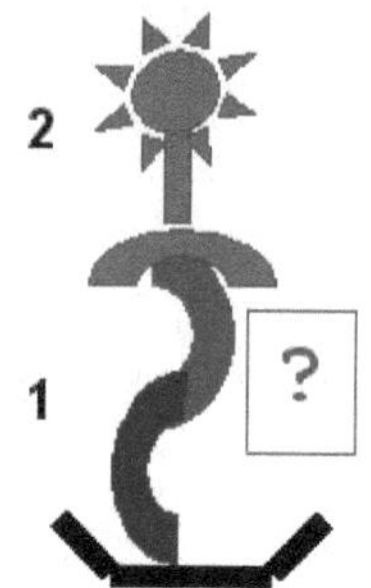

Figure 51: Hybridisation

1   - Hybrid formation
**viral nucleic acid from the sample (DNA or RNA)/probe specific to the viral
genome sought,** on various **media:** membrane, microplate, tube .
2   - Enzyme immunodetection of
the molecular hybrid :
by **anti-double nucleic acid antibodies
strand,** or cold marking of the probe .
followed by **colorimetric,** fluorescent or chemiluminescent **detection.**
4.4.1. Hybridization with signal amplification
-    Principle of the DNA branch :
The viral nucleic acid (sample) is
-    adsorbed onto a support using capture probes, then
-    detected by branched probes on which
-    hybridize revelation probes.
Luminescent molecules are present on the detection probes; the luminescent
signal is proportional to the number of adsorbed viral nucleic acid molecules.
The simultaneous use of numerous branched probes, complementary to different
regions of the target viral genome, enables the detection of genetically highly
variable viruses (HIV, HCV). The large number of luminescent molecules
involved means that the detection threshold can be lowered, without any risk of
contamination (unlike PCR, there is no amplification of the nucleic acid).

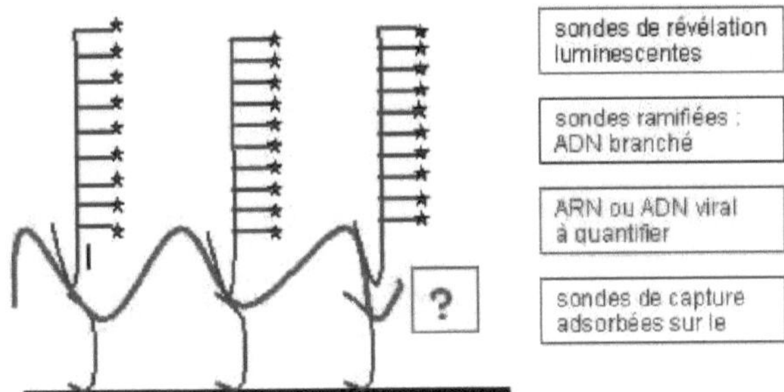

Figure 52: Hybridisation with amplification

Common applications : ARN HIV-1 , ARN HVC, ADN HVB

4.4.2. Hybridization on membranes and chips

- Principle of the different methods:

1- hybridization to molecular probes fixed on a support and specific to either a wild-type virus or a mutated virus, then

2- enzyme-linked immunosorbent assay.

The probes are present on strips, rather like antigens fixed on immunoblot strips (e.g. HVC genotyping, HVB mutations).

Microarrays' will also enable the detection of mutated viral nucleic acids. Large numbers of probes are adsorbed onto silica supports, representing a kind of combination of computer and molecular techniques. An application will soon be available for the detection of HIV mutations associated with resistance to anti-retroviral drugs.

Genotyping of HVC by membrane hybridisation after PCR: the bands are interpreted using a template supplied by the manufacturer. In this case, the patient carries genotype 1 CVH, subtype b, which is more resistant to anti-CVH treatment.

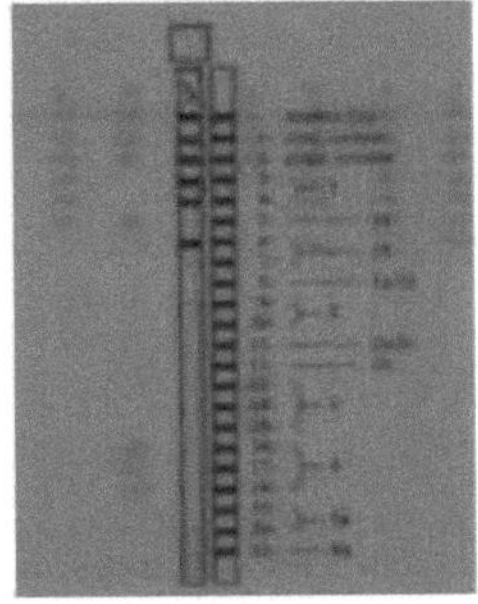

Figure 53: Genotyping

103

4.4.3. Gene amplification
i Principle of PCR :

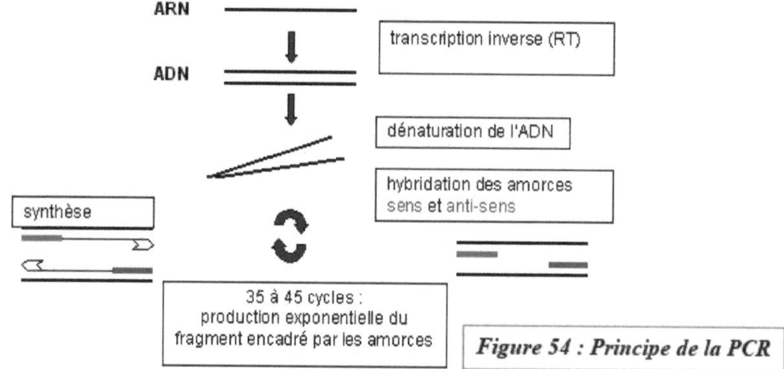

Figure 54 : Principe de la PCR

Figure 54: Principle of PCR

4- DNA or RNA is extracted from a sample using a variety of methods. In routine use, extraction is often achieved by adsorption of nucleic acids (negatively charged) onto positively charged silica particles, followed by elution.

5- Gene amplification, or PCR (polymerase chain reaction), enables a limited portion of the viral genome to be copied, using primers and the action of a DNA polymerase.

6- The primers are designed to hybridise to the viral genome, one on the sense strand, the other on the anti-sense strand. Primers can therefore only be chosen for a virus whose nucleotide sequence is known; in addition, the primers must be specific to the virus, and not hybridise to any genome other than the one under investigation.

7- The DNA polymerase "hooks" onto the two primers and synthesises the complementary DNA from the viral genome that serves as its model; synthesis is made possible by adding dATP, dCTP, dGTP and dTTP to the reaction medium. Each PCR cycle comprises a DNA denaturation step, a primer hybridization step and then a DNA synthesis step. These steps are carried out in a thermocycler by three successive temperature changes (e.g. 94°C, 60°C, 72°C).

8- To amplify the RNA, a preliminary step of reverse transcription into DNA is essential before PCR; this is carried out in vitro using a reverse transcriptase (of avian or murine origin). All the above reagents are commercially available.

9- After gene amplification, the amplified nucleic acids need to be detected, either by agarose gel electrophoresis or, more commonly, by post-PCR molecular hybridization.

10- Common applications : HVC, HVB, CMV, HHV, neurotropic viruses, parvovirus B19, ...

Four types of conventional thermocycler: *a thermocycler is a programmable (temperature and time) dry water bath.*

Figure 55: PCR equipment

Γagarose gel after PCR and electrophoresis: *The gel is photographed under UV light. Positive signals appear as clear bands (arrows). The result can only be interpreted if there are positive controls (the reaction has worked) and a negative control (there is no contamination).*

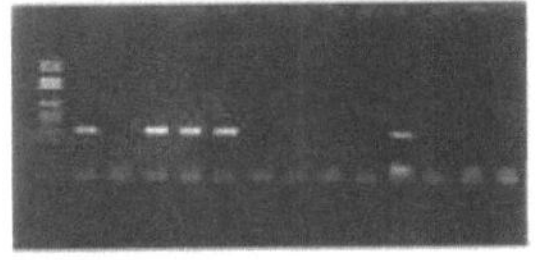

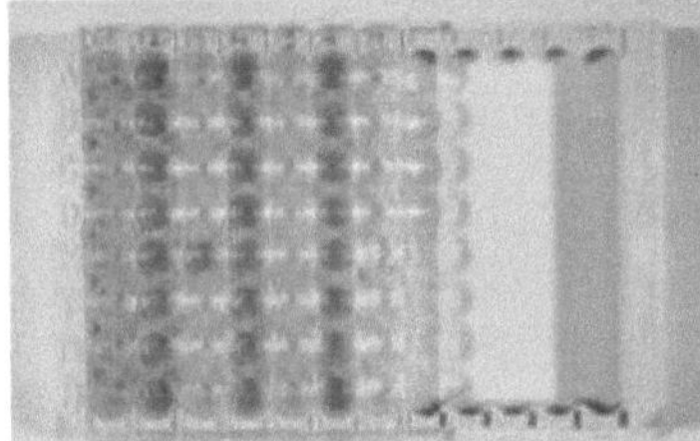

Figures 56-58: PCR results

Aspect of a post-PCR molecular hybridization microplate: *coloured wells correspond to positive signals.*

4.4.4. Nucleotide sequencing

The Sanger method is used for routine sequencing. It is a PCR in which fluorochrome-labelled dideoxynucleotides (ddATP, ddCTP, ddGTP, ddTTP) are added to the usual reaction mixture. The polymerase synthetises the DNA using either deoxynucleotides or dideoxynucleotides at random. Each time a dideoxynucleotide is incorporated into the DNA chain, chain elongation stops, as these molecules do not allow the next phospho-diester bond. The sequence reaction therefore results in the production of DNA fragments of all possible sizes (for example, for a recopied DNA fragment of 200 nucleotides, from 1 to 200). Automatic" sequencers automatically detect the fluorochrome on these DNA fragments using electrophoresis. This is carried out either in a vertical acrylamide gel, or in polymer-filled Sicilian capillaries.

Example of an automatic sequencer: the electrophoresis capillaries and the reading system are located in the device on the left. Interpretation is performed using specialised software on a PC connected directly to the sequencer.

Figure 59: Automatic sequencer

Analysis of a sequence reaction after reading on an automatic sequencer : the two DNA strands, sense and antisense, have been sequenced (two electrophoregrams) and the sequences are compared with each other and with a known reference sequence. The 'consensus' represents the final sequence, obtained after human correction of ambiguities not resolved by the automatic sequencer.

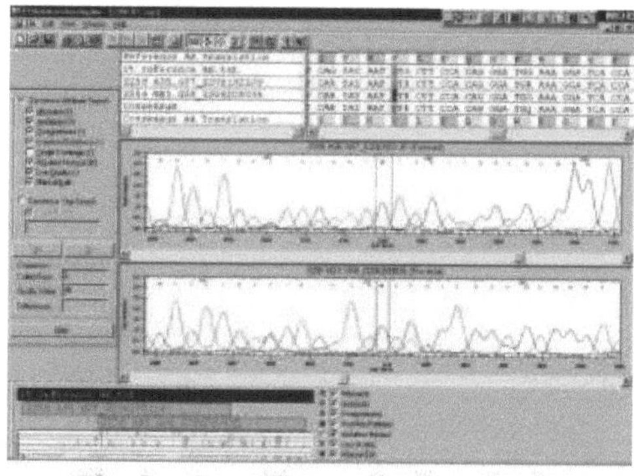

Figure 60: **Analysis of an ae sequence reaction**

4.4.5. Quantification
-   **Branch DNA** is a quantification method.
-   **Competitive PCR with internal standard**
- **Principle of competitive PCR quantification** :

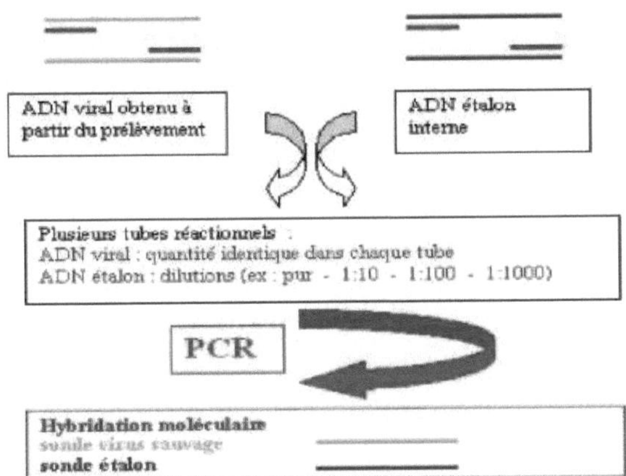

Figure 61: **Quantification by competitive PCR**

Zone of equivalence: optical density of **wild virus** = optical density of **standard.**

Various PCR kits on the market are based on this principle. A standard is introduced into the samples to be analysed, before the nucleic acids are extracted (internal standard). This standard is a nucleic acid whose extremities are identical to those of the wild-type virus sequence, which means that the same pair of primers can be used. However, between the ends to which the primers hybridise, the nucleotide sequence differs from that of the wild-type virus, making it possible to distinguish the two PCR products, either by electrophoresis (different size) or by hybridisation after PCR (different specific probes).

Each sample is tested in several tubes. In each tube, the sample is always present in the same quantity, but the standard is distributed in different concentrations (dilutions of the standard). During PCR, the primers hybridise either to the wild-type sequence (virus in the sample) or to the standard: there is therefore competition (competitive PCR). The primers hybridise more readily to DNA present in greater quantity, so it is amplified more. In a tube, the same amplification is obtained for both matrices (equivalence zone): as the concentration of standard is known, this makes it possible to deduce the quantity of molecules in the sample.

Real-time chain polymerisation reaction

Real-time PCR methods measure the amount of DNA molecules made during the early exponential phase of the PCR reaction. Schematically, the DNA made during PCR is quantified by incorporation of a fluorescent molecule or hybridization of two specific fluorescence-emitting probes. These particular thermal cyclers are equipped for continuous (real-time) reading of the fluorescence emitted.

Example of a real-time PCR device:

***Figure 62:* Real-time PCR apparatus**

PCR and detection by hybridisation are carried out at the same time in the same tube

EBV quantification curves in whole blood using real-time PCR: *the top curves show the amplification of DNA from positive controls at different concentrations (external standard curve). The lower curve represents the slope.*

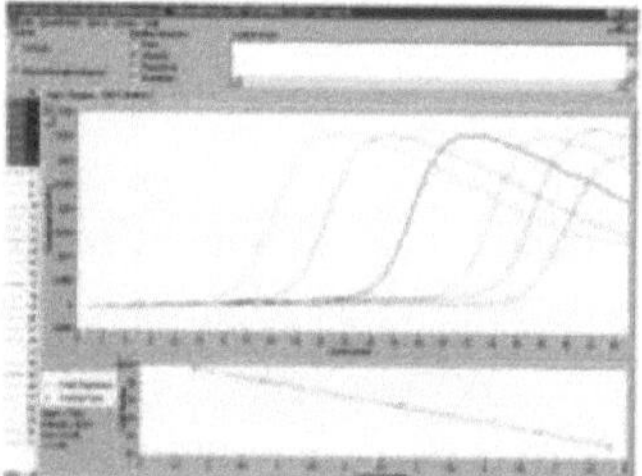

***Figure 63:* PCR** *quantification curves*

Indications for molecular diagnostic methods :

-Qualitative techniques: to show the presence of a virus, particularly when it cannot be cultured (HVC), or when the biological fluid is not suitable for culture (CSF).

-    **Quantitative techniques**: monitoring patients undergoing anti-viral treatment (HIV, HCV, CMV, HBV).

-    **Sequencing**: detection of nucleotide mutations

-    associated with resistance to anti-virals (HIV, HVB),

-    characteristics of virus genotypes (HVC, HVB).

- Interests :

-    Remarkable sensitivity

-    Automation possible

- Limits :

-    Contamination risks for PCR techniques.

-    No evidence that the virus detected is infectious.

-    Phenotypic characterisation of the strain impossible.

-    Still relatively expensive.

1. What is a virus?
2. Give a brief history of virology
3. Make the structure of a naked virus and an envelope virus
4. Why do people say that viruses are very small?
5. What do anti-receptors represent in a virus?
6. Give the general characteristics of viruses
7. What differences do you see between helicoidally symmetrical viruses, icosahedral viruses and viruses with complex architecture?
8. Demonstrate that viruses replicate
9. What are the limits of the LHT System Classification and the David Baltimore Classification?
10. Using specific examples, demonstrate that viruses are specific to a particular disease.
11. Using clearly labelled diagrams, explain the development cycle of naked viruses, enveloped viruses and retroviruses (HIV or HBV).
12. Using the compartmentalisation of eukaryotic cells, explain the development cycle of DNA and RNA viruses.
13. Explain the genetic variations in plant viruses.
14. Is it possible to vaccinate plants? If so, explain the process.
15. Using specific examples, explain the methods used to combat the following viral diseases: Poliomyelitis, Ebola, Convid-19, Influenza, Rabies, AIDS and Hepatitis B.
16. What methods are used to diagnose the diseases listed in 15?
17. What methods are used to diagnose illnesses in your country?
18. Which of the diagnostic methods you are familiar with seems to be the most effective? Justify your answer.

NOBEL PRIZE WINNERS IN MEDICINE, PHYSIOLOGY OR CHEMISTRY
WHOSE WORK STEMS FROM VIROLOGY OR WHO HAVE HAD AN INFLUENCE ON VIROLOGY

In 1946 [chemistry] J.H Northrop, W.M Stanley, J.P Sumner: preparation in the pure state of enzymes and viral proteins (J.H.N., W.M.S), discovery of enzyme crystallisàtion (JPS).

In 1951 [medicine] M. Theiler: vaccine against yellow fever.1954 [medicine] J.F Enders, T.H. Weller, F.C. Robbins: growth of poliomyelitis virus in cell culture.

In 1958 [medicine-physiology] G.Beadle, E.L. Tatum and J.Lederberg: discovery of the role of genes in protein synthesis (G.B and E.L.T) and the discovery of recombination and the organisation of genetic material in bacteria (J.L).

In 1962 [medecine] F.H.C. Crick, J.D. Watson, M.H.F. Wilkins: structure of the DNA double helix.

In 1965 [medicine] F.Jacob, A.Lwoff, J.Monod: Genetic control of enzyme and virus (bacteriophage) synthesis.

In 1966 [medicine] P. Rous: discovery of oncogenic viruses.

In 1969 [medecine] M.Delbruck, A.D. Hershey, S.E. Luria: Mechanisms of replication and genetic structure of viruses.

In 1975 [medecine] D Baltimore, R Dulbecco, HM Temin: Interactions between oncogenic viruses and the genetic material of cells.

In 1976 [medicine] B.S. Blumberg, D.C. Gajdusek: New mechanisms of the origin and dissemination of infectious diseases (hepatitis B virus and Kuru).

In 1978 [medicine] W. Arber, D. Nathans, H.O. Smith: The discovery of restriction enzymes and their application in molecular genetics.

In 1980 [chemistry] P. Berg, W. Gilbert, F. Sanger: fundamental studies of the biochemistry of nucleic acids with particular attention to recombinant DNA (PB) and contribution to the determination of nucleic acid base sequences (WG, FS).

In 1982 [chemistry] A. Klug: Elucidation of the structure of important nucleoproteins in biology, by electron microscopy and crystallography.

In 1989 [medecine] J.M. Bishop, H.E. Varmus: Cellular origin of retroviral oncogenes.

In 1993 [medicine] R.J. Roberts, P.A Sharp: Genes mosaics (work on viruses, adeno and SV40).

In 1993 [chemistry] KB Mullis, M.Smith: PCR (KBM) and directed mutagenesis (MS).

In 1997 [medecine] S.B. Prusiner: prions, a new biological principle of infection.

In 2006 [medecine] A. Fire and G. Mello: RNA interference - gene silencing by double-stranded RNA.

In 2008 [medicine] Harald zur Hauzen: work on cervical cancer caused by the papillomavirus. Fran^oise Barre-Sinoussi, Luc Montagnier: discovery of the AIDS virus.

In 2020 Michael Houghton [biology], Harvey J. Alter [medicine] and Charles M. Rice [medicine]: For the discovery of the hepatitis C virus.

In 2023 Katalin Kariko [Biology] and Drew Weissman [Biochemistry]: For their discoveries concerning modifications to nucleoside bases that have enabled the development of effective mRNA vaccines against Covid-19.

BIBLIOGRAPHY

David M. Knipe, Peter M. Howley, Wolters Kluwer, Fields Virology, Fifth edition, - Lippincott Williams & Wilkins : Virology 2 volumes, 2007.

Nigel J. Dimmock, Abdrew J. easton, Keith n. leppard: Introduction to Modern Virology, sixth edition, Blackwell Publishing, 2007.

Huraux J.-M., nicolas J.-C., agut H., Peigue-lafeuille H. (ed) 2003, Traite de Virologie Medicale, estem De Boeck Diffusion, Paris, France. (UCL-bibliotheque de Medecine).

Fenner F, Henderson DA, Arita I, Jezek Z, Ladnyi ID. Smallpox and its eradication, World Health Organization, geneva, 1988, pp 1460. (Reference UCl : 10066479, location : bibliotheque de medecine)

Pasteur l, Chamberland C, roux e. (1884). Physiologie experimentale: nouvelle communication sur la rage. C. R. Acad. Sci.98: 457-63.

Rous P. (1911). A sarcoma of the fowl transmissible by an agent separable from tumor cells. J Exp Med, 13: 397-9.

Watson JD. The double helix. a personal account of the discovery of the structure of DNA, new american library, New-York, 1968 (reference UCl: 500087107, location: bibliotheque des sciences exactes).

Saiki R.K, Scharf S., Faloona F., Mullis K.B, Horn G.T, Erlich HA, Arnheim N. (1985). Enzymatic amplification of beta-globin genomic sequences and restriction site analysis for diagnosis of sickle cell anemia. Science, 230 :1350-4. (bibliotheques uCl: bibliotheque de medecine et bibliotheque des sciences exactes et bibliotheque de Hemptinne)

temin HM, Mitzutani s. (1970). RNA-dependant DNA polymerase in virions of Rous sarcoma virus. Nature, 226 : 1211-1213.(uCl libraries : library of medicine and library of exact sciences)

Baltimore D. (1970). RNA-dependent DNA polymerase in virions of RNA tumour viruses. Nature, 226 :1209-11. (uCl libraries: library of medicine and library of exact sciences).

Fauquet C.M., Mayo, M.a., Maniloff, J., Desselberger, U., Ball, L.A. 2005. Virus taxonomy. 8th report of the International committee on taxonomy of viruses, elsevier academic Press, Oxford, uK, 1259 pp.

Buchen-Osmond C. (2003). The Universal Virus Database ICTVdB. Computing in science and engineering 5 (3), 16-25.

ICTV website: http://www.iums.org/comcofs/comcofs-virology.html, http://www.mcb.uct.ac.za/ictv/ICtv.html, http://www.danforthcenter.org/iltab/ictvnet/asp/_MainPage.asp

ICTV-DB website: http://www.ncbi.nlm.nih.gov/ICtvdb/

Principles of Molecular Virology. third edition, alan J. Cann, academic Press

Barbot J., "Agir sur les essais therapeutiques. L'experience des associations de lutte contre le sida en France", *Revue d'Epidemiologie et Sante Publique,* 1998, 46, p. 305-315.

Barbot J., *Les malades en mouvements. La medecine et la science a l'epreuve du sida,* Paris, Editions Balland, 2002.

Barbot J., Dodier ?, " L'emergence d'un tiers public dans la relation malade-medecin. L'exemple de l'epidemie a vih", *Sciences Sociales et Sante,* 2000, 18, 1, p. 75-119.

Decreased Mortality of COVID-19 with Renin-Angiotensin-Aldosterone System Inhibitors Therapy in Patients with Hypertension: A Meta-Analysis [HYPERTENSION] 27/05/2020.

Out-of-hospital cardiac arrest during the COVID-19 pandemic in Paris, France: a population-based, observational study [THE LANCET] 27/05/2020.

PCR Spotlight: COVID-19 - How to restart an elective programme in interventional cardiology [PCR ONLINE] 26/05/2020

Advocacy in Action: Newest CMS COVID-19 Rule Includes Telehealth Measures Requested By ACC [ACC] 22/05/2020.

PUBLIC HEALTH AGENCY OF CANADA. "Summary of recommendations for preventing the contraction of viral hepatitis while travelling", [Online], Canadian Communicable Disease Report, vol. 40, no. 13, July 10, 2014, pp. 310-314.

PUBLIC HEALTH AGENCY OF CANADA. Updated recommendations for the use of hepatitis A vaccine , 2016, 43 p.

GILCA, Vladimir, and Nicole BOULIANNE. *Avis n° VHA/2009/001 : Utilisation du vaccin contre le VHA chez les enfants de moins de 1 an*, Institut national de sante publique du Quebec, 4 p. [Avis interne examine et approuve par le CIQ le 12 mars 2009].

"Fievre de Lassa", on the Caisse des Frangais de l'Etranger website (consulted on 26 December 2018)

(en) Werner Dietrich, *Biological Resources and Migration*, Springer, 2004, 363 p. (ISBN 978-3540214700)

Institut Pasteur (pasteur.fr): "Lassa fever", March 2008 (consulted on 29 December 2011)

WHO 2005: "Lassa fever" April 2005 (consulted on 29 December 2011) Ogbu O, Ajuluchukwu E, Uneke CJ, *Lassa fever in West African sub-region: an overview*, vol. 44, 2007, 1-11 p.

(en) Antonio Tenorio, W. Ian Lipkin, Juan E. Echevarria and Stephen K. Hutchison, "Discovery of an Ebolavirus-Like Filovirus in Europe", *PLOS Pathogens*, vol. 7, n° 10, 20 October 2011, e1002304 (ISSN 1553-7374, PMID

22039362, PMCID PMC3197594, DOI 10.1371/journal.ppat.1002304, read online [archive], accessed 17 Aug 2019)

(en) Zheng-Li Shi, Lin-Fa Wang, Yun-Zhi Zhang and Ren-Di Jiang, "Characterization of a filovirus (Mengla virus) from Rousettus bats in China", *Nature Microbiology*, vol. 4, n° 3, March 2019, pp. 390-395 (ISSN 2058-5276, DOI 10.1038/s41564-018-0328-y, accessed 17 August 2019)

ICTV. International Committee on Taxonomy of Viruses. Taxonomy history. Published on the Internet https://talk.ictvonline.org/., consulted on 5 February 2021 (en) "Virus Taxonomy: 2018b Release" [archive], ICTV, July 2018 (consulted on 4 July 2019).

(en) "Biosafety in Microbiological and Biomedical Laboratories, 5th Edition" [archive], CDC, December 2009 (consulted on 25 July 2019).

(en-US) Mariette F. Ducatez, Claire Pelletier and Gilles Meyer, "Influenza D Virus in Cattle, France, 2011-2014", *Emerging Infectious Diseases*, vol. 21, no. 2, Feb. 2015 (DOI 10.3201/eid2102.141449, read online [archive], accessed 15 Feb. 2018)

Corinne Thebault Le 21 fevrier 2003 a 00h00, " Le virus de la grippe sous l'influence d'El Nino ! " [archive], on leparisien.fr, 20 february 2003 (consulted on 14 december 2021)

"La Nina favours the appearance of new flu viruses" [archive], at www.20minutes.fr (consulted on 14 December 2021)

R. Cattaneo and M. McChesney, *Measles Virus*, in B. W. J. Mahy (Ed.), Encyclopedia of virology (3rd ed., Vol. 3). Amsterdam, Boston, Academic Press, 2008, p. 285-288

Erling Norrby, *Measles*, in B. N. Fields, Virology, New York, Raven Press, 1985, pp. 1305-1309

Thierry Borrel, *Les virus*, Paris, Nathan, coll. Sciences 128, 1996

Helene Valentin, Branka Horvat, Pierre Rivailler and Chantai Rabourdin-Combe, *Role of the CD46 molecule in measles virus infection*, Annales de l'Institut Pasteur/Actualites, Volume 8, Issue 2, July-September 1997, Pages 181-184.

Pr. M.E. Lafon (Virology Laboratory, Faculty of Medicine, Victor Segalen University, Bordeaux).

A. Mammette, Collection AZAY: " Virologie medicale " Presses Universitaires de Lyon. 2002, 80 Boulevard de la Croix Rousse, BP 4371, 69242 LYON 04

Baltimore D, "Expression of animal virus genomes", *Bacteriol Rev*, vol. 35, n° 3, 1971, p. 235-41 (PMID 4329869, read online [archive])

Professor Colimon, "Classification modifiee de Baltimore selon la stratégie de replication des virus [archive]", Universite de Rennes I - Departement de virologie, 16 October 2001 (consulted on 22 November 2023).

Consolidation questions and topics

1. After giving the history of virology, why are viruses said to be very small?
2. What is the structure of a naked virus?
3. What is the structure of an envelope virus?
4. What do you know about how enveloped viruses enter a host cell?
5. What do you know about how naked viruses enter a host cell?
6. What is the difference between the viral matrix and the cell receptor?
7. What does an anti-receptor mean in a virus?
8. Using a well-annotated diagram, draw up the structure of the Flu Virus.

Course summary

This textbook on General and Special Virology is intended for Licence 2 students in Medical Biology, Medicine, Pharmacy, Odontostomatology and Biology. It comprises two main parts. The first part deals with the introduction to virology, definition of viruses, history of virology, general characteristics of viruses, specificity of viruses, dependence of viruses on the metabolism of host cells, structure of the viral particle (genome, capsid, lipid membrane, envelope, matrix, anti-receptors, naked viruses and enveloped viruses), viruses with helical symmetry, viruses with icosahedral symmetry and viruses with complex architecture.

It also deals with the criteria for distinguishing viruses, viral taxonomy, classification of viruses according to the LHT System and that of David Baltimore, the concept of viral species, specificity of viral taxonomy, collection and conservation of viruses, the development cycle according to the nature of the viruses (attachment, expression and replication, virus assembly and exit), the special case of plant viruses, the compartmentalisation of the eukaryotic cell and the stages of replication, transcription and maturation of messenger RNAs (mRNAs) taking place in the nucleus. Translation of the mRNAs into proteins takes place in the cytoplasm. The proteins produced are then sent to the compartment where they are to exert their action: nucleus, cytoplasm, endoplasmic reticulum, lysosomes, mitochondria, membrane, etc.

The second part of this manual deals in detail with the special virology of nine viruses and the diseases they cause in man. These are HIV and AIDS, Poliovirus and Poliomyelitis, SARS Covid 2 and Covid-19, Hepatitis B Virus and Hepatitis B, Lassa Virus and Lassa Fever, Ebola Virus and Ebola Virus Disease, Marburg Virus and Marburg Virus Disease, Influenza Virus and Influenza, and Measles Virus and Measles.

The final part of the book deals with the methods used in laboratories to diagnose infectious viral diseases, and the different stages involved in each of the methods studied.

Dr Taliby Dos CAMARA is a lecturer and researcher in General Microbiology, Medical Microbiology, Medical Mycology, Virology, Medical Bacteriology, Biosecurity and Research Methodology at the Gamal Abdel Nasser University in Conakry (UGANC), the Mahatma Gandhi University (UMG) and the Institut Recherche en Biologie Appliquée de Gurnee (IRBAG). He is registered on the I aptitude list of the Centre Africain et Malgache pour l'Enseignement Superieur (CAMES).

Printed by Books on Demand GmbH, Norderstedt / Germany